T0256753

Measuring Innovation Everywhere

To
Adelaide, Byron and Frances

Measuring Innovation Everywhere

The Challenge of Better Policy, Learning, Evaluation and Monitoring

Fred Gault

Professorial Fellow, UNU-MERIT, the Netherlands, Professor Extraordinary, Tshwane University of Technology (TUT) and Visiting Professor, DST/NRF/ Newton Fund Trilateral Research Chair in Transformative Innovation, the 4th Industrial Revolution and Sustainable Development, University of Johannesburg, South Africa

Edward Elgar
PUBLISHING

Cheltenham, UK • Northampton, MA, USA

Published by
Edward Elgar Publishing Limited
The Lypiatts
15 Lansdown Road
Cheltenham
Glos GL50 2JA
UK

Edward Elgar Publishing, Inc.
William Pratt House
9 Dewey Court
Northampton
Massachusetts 01060
USA

A catalogue record for this book
is available from the British Library

Library of Congress Control Number: 2020938629

This book is available electronically in the **Elgar**online
Economics subject collection
DOI 10.4337/9781789904567

MIX
Paper from
responsible sources
FSC
www.fsc.org   FSC® C013056

ISBN 978 1 78990 455 0 (cased)
ISBN 978 1 78990 456 7 (eBook)

Printed and bound in Great Britain by TJ International Ltd, Padstow, Cornwall

Contents

List of tables   vii
Preface   viii
Acknowledgements   x
List of abbreviations   xiv

PART I   INTRODUCTION

1   Introduction to 'Innovation Everywhere'   2

2   Innovation systems   7

PART II   INNOVATION POLICY

3   Innovation policy prior to 2020   15

4   Monitoring and evaluation of innovation policy   29

5   Developing innovation policy   40

PART III   MEASURING INNOVATION

6   Defining innovation for measurement purposes   47

7   Measuring innovation in all economic sectors   58

8   Measuring innovation across economic sectors   69

PART IV   WHERE NEXT?

9   Innovation and future challenges   79

10   Innovation, measurement and policy   91

11   Conclusion   102

References 106
Index 117

Tables

2.1	Key activities in innovation systems in Borrás and Edquist (2019: 25)	11
3.1	More firms innovate than do R&D	20
3.2	Reporting innovation in the business sector	21
9.1	SDG sub-goals related to the corresponding FDES statistics	86
10.1	Business characteristics and innovation	93

Preface

Since the 2010 book (Gault 2010), things have changed. The issues in 2010 were the recovery from the financial crisis and the need for jobs and growth. There was some acknowledgement of sustainability of changes introduced to the economy and the society, and of other global challenges such as climate change and energy. Innovation, seen as a way of delivering jobs and growth, was being measured and entered into official statistics. The thought was that if innovation could be measured, innovation policy could be developed, implemented and monitored. This could lead to evaluation of innovation policy and, perhaps, to policy learning. A decade ago, official statistics for innovation were collected from the business sector, but there was discussion of public sector innovation, even of an equivalent to the *Oslo Manual* for the public sector. These discussions continue, including the need for a manual to support the measurement of household sector innovation.

The difficulty with producing manuals for the measurement of innovation outside the business sector was the definition. So far, it required a new or significantly improved product to be introduced to the market for it to be an innovation (OECD 1992; OECD/Eurostat 1997, 2005). This was a problem for the public sector and the household sector. In 2018, this changed, with the release of the fourth edition of the *Oslo Manual* (OECD/Eurostat 2018). This is the first major change since 2010.

The general definition of innovation in the fourth edition of the *Oslo Manual* requires that the product is made available to potential users. Of course, introducing the product on the market is one way of making it available to potential users, but not the only one. This book examines the case when products (goods or services) are made available to potential users at a zero price. This happens in the digital economy.

The second major change since 2010 is the size and the rapid growth of the digital economy and how it affects innovation and its statistical measurement. Zero price products are an issue, not just for understanding innovation but also for the System of National Accounts.

The third major change is the growing importance of restricted innovation. This is innovation as defined in the *Oslo Manual*, but with an added constraint that the innovation be green, inclusive, sustainable, or any other restriction of policy interest. The importance of this is that all of the basic definitional issues in the *Oslo Manual* are unchanged and the statistical measurement of innovation can take place. Once that is done, and the innovators are identified, additional questions can follow as to whether the innovation exhibits the restriction, or not. For this to happen, the restriction has to be well defined, but that is a measurement issue that can be overcome. Time is also an issue. If the policy is to improve inclusion, there must be a baseline measure followed by subsequent measures to demonstrate that inclusion has increased. The book examines these issues which were not considered in 2010.

The book also examines innovation in the informal economy and considers the difficulty of measuring 'social innovation' and it ends with an agenda for work to follow. Some of the recommendations for future work are unchanged from 2010, but not all. Understanding innovation, its measurement, and the impact of innovation policy is a growing subject and an important one.

Acknowledgements

This book is, in part, an update of Gault (2010), taking account of the issues raised there, and how they have changed over a decade, or not. Given that the book is about statistical measurement of innovation, and the influence of the resulting indicators on innovation policy, the biggest change has been in the definition of innovation.

The fourth edition of the *Oslo Manual* (OECD/Eurostat 2018) introduced a general definition of innovation applicable in all economic sectors. This was a big step and I was able to participate in the discussions as a result of a grant from the Government of Norway facilitated by Sven Olav Nås, then the Chair of the Organisation for Economic Co-operation and Development (OECD) Working Party of National Experts on Science and Technology Indicators (NESTI), and managed by Fernando Galinda-Rueda at the OECD.

The discussion of a general definition goes back some time. In the course of a Finnish project on consumer innovation, led by Jari Kuusisto and Eric von Hippel, the fact that 'user innovators' that shared their new or significantly improved product with peer groups, or communities of practice, were not innovators was a concern. The issue was that these 'user innovators' did not connect to the market. I am indebted for the spirited discussion that followed. I proposed a way to resolve this and the solution was to change 'introduced on the market' to 'made available to potential users' Gault (2012). In the course of examining the consequence of using 'made available to potential users' it became clear that it could be applied to the public sector. The results of the Nordic MEPIN project on public sector innovation were examined with valuable input from Carter Bloch. This work on definitions would not have gone ahead had it not been for the contributions of Carter Bloch, Eric von Hippel and Jari Kuusisto. The next step was to deal with all economic sectors in a coherent manner.

At the same time as these discussions were going on, I was editing the *Handbook of Innovation Indicators and Measurement* (Gault 2013). The contributors to that book provided ideas for the revision of Gault (2010), once the debate on the general definition of innovation was resolved.

I am indebted to Erika Rost, a former Vice-Chair of NESTI, for her suggestions for structuring an updated version of Gault (2010), incorporating findings from the *Handbook* and other sources.

In 2015 the revision of the *Oslo Manual* began and a UNU-MERIT Working Paper (Gault 2015) was made available to members of NESTI, past and present, who were engaged in the revision. The discussion of general definitions by NESTI delegates and experts contributed significantly to the understanding of a general definition which was published in 2018 (Gault 2018). The general definition in that paper was not exactly the same as the one in the fourth edition of the *Oslo Manual*, but the *Oslo Manual* achieved the objective of having a general definition of innovation. All of the NESTI delegates contributed to that.

Outside of NESTI, the concept of the general definition was attracting interest. There were discussions among participants in the African Science, Technology and Innovation Indicators (ASTII) initiative led by Aggrey Ambali, and the African Observatory on Science, Technology and Innovation (AOSTI) led by Philippe Mawoko with Almamy Konté. The author has had a long-standing relationship with the Centre for Science, Technology and Innovation Indicators (CeSTII) initiated by Michael Kahn in 2002 and continued with Glenda Kruss. It has always been a pleasure to work with the CeSTII team and the Human Sciences Research Council. I thank Rasigan Maharjh for my association with the Tshwane University of Technology (TUT) in Pretoria and the Institute for Economic Research on Innovation (IERI). More recently, I thank Erika Kraemer-Mbula, who holds the DST/NRF/Newton Fund Trilateral Research Chair in Transformative Innovation, the 4th Industrial Revolution and Sustainable Development at the University of Johannesburg, for my involvement with that group.

Well before the decision to revise the *Oslo Manual* in 2015, the US National Research Council convened, in 2011, a Panel to review the question of *Capturing Change in Science, Technology, and Innovation: Improving Indicators to Inform Policy*. I participated at the invitation of the co-chairs, Robert E. Litan and Andrew W. Wyckoff, and the report (National Research Council 2014), remains relevant to this day. Kaye Husbands Fealing managed the Panel and more recently provided insights on the future of the NSF Science of Science and Innovation Policy (SciSIP) programme. As part of the US ongoing work on indicators, John Jankowski, the current Chair of NESTI, invited me to contribute to an NCSES/CNSTAT event, chaired by Scott Stern, on *Advancing Concepts and Models of Innovation Activity and STI Indicator Systems* at the National Academies of Sciences, 19–20 May 2016 in Washington. As with the report just cited (National Research Council 2014), the resulting report (National Academies of Sciences, Engineering and Medicine 2017) considered different ways of measuring innovation.

The OECD Blue Sky Forum in 2016 contributed to the discussion that was part of the revision of the *Oslo Manual*. I thank Ward Ziarko, another former Chair of NESTI, and the organisers of the Forum, for their invitation to take part.

Over a decade, I benefitted from discussions with Bengt-Åke Lundvall, Charles Edquist and John Marburger on different aspects of innovation indicators and their use. The Science Policy Research Unit (SPRU) has played a role from the beginning of this project through discussions with SPRU researchers. I particularly thank Ben Martin.

The book is about innovation policy as well as statistical measurement and I am indebted to Manuel Heitor, Minister of Science, Technology and Higher Education in Portugal for discussions organised by Joana Mendonça and Giorgio Sirilli, another former Chair of NESTI.

Lili Wang, of United Nations University – Maastricht Economic and social Research and training centre on Innovation and Technology, the Netherlands (UNU-MERIT), has promoted discussion of indicators and their development, including the impact of a general definition of innovation, which contributed to this work.

While there have been many meetings over the last decade looking at definitions of innovation, there was another activity growing rapidly which was the digital economy. This raised measurement questions addressed by the OECD in *Measuring the Digital Transformation: A Roadmap for the Future* (OECD 2019a). It was presented to me by Alessandra Colecchia with the suggestion that I read it. The text will demonstrate that I did.

Another topic which overlapped with the general definition of innovation was the *Maastricht Manual on Measuring Eco-Innovation for a Green Economy*. This was given to me for review and it is a very ambitious project, led by René Kemp, that should make statisticians and policy people think. It is discussed in Chapter 8.

Anthony Arundel and Carter Bloch provided useful comments on earlier work as did students of a course I give occasionally at UNU-MERIT. It has been a privilege to have been a Professorial Fellow at UNU-MERIT for over a decade and I am indebted to Luc Soete for appointing me in 2009 and Bart Verspagen for his leadership of

UNU-MERIT for the last eight years. I was supported by Eveline in de Braek and Marc Vleugels in various projects leading to this book.

Books are always a challenge and I wish to thank Edward Elgar Publishing for its corporate support, Matt Pitman for being both encouraging and accommodating and Stephanie Hartley for her guidance.

While there have been many inputs and useful comments, the final text and any errors are the responsibility of the author.

Abbreviations

AI	Artificial intelligence
AIO	African Innovation Outlook
AOSTI	African Observatory of Science, Technology and Innovation
ASTII	African Science, Technology and Innovation Indicators
AU	African Union
AUDA	African Union Development Agency
BMWi	Federal Ministry for Economic Affairs and Energy
BRDIS	Business R&D and Innovation Survey (US)
CEC	Commission for the European Communities
CIS	Community Innovation Survey (EU)
CNSTAT	Committee on National Statistics
CYTED	Programa Iberoamericano de Ciencia y Technological para el Desarrollo
DARPA	Defense Advanced Research Projects Agency
DK	Denmark
DST/NRF	Department of Science and Technology/National Research Foundation
EC	European Commission or European Community
Eco	Ecological
EFI	Commission of Experts for Research and Innovation
EIC	European Innovation Council
EIS	European Innovation Scoreboard
EPSIS	European Public Sector Innovation Scoreboard
EU	European Union
Eurostat	Statistical Office of the European Communities

EUSPRI	European Forum for the Studies of Politics of Research and Innovation
FDES	Framework for the Development of Environmental Statistics
GDP	Gross domestic product
GERD	Gross domestic expenditure on research and development
GII	Global Innovation Index (INSEAD/WIPO)
GTIPA	Global Trade and Innovation Policy Alliance
HTGF	High-Tech Start-up Fund
ICLS	International Conference of Labour Statisticians
ICT	Information and communication technology
IERI	Institute for Economic Research on Innovation
ILO	International Labour Organization
IMF	International Monetary Fund
INSEAD	Institut Européen d'Administration des Affaires (France)
IP	Intellectual property
IPP	OECD-World Bank Innovation Policy Platform
ISI	International Statistical Institute
ISIC	International Standard Industrial Classification
ISO	International Organization for Standardization
ISOC-SE	Internet Society Sweden
ITU	International Telecommunications Union
JRC	Joint Research Centre
KAU	Kind-of-activity unit
MDGs	Millennium Development Goals
MEPIN	Measuring Public sector Innovation in the Nordic countries
MERIT	Maastricht University's Economic Research Institute on Innovation and Technology (now: UNU-MERIT)
MIoIR	Manchester Institute of Innovation Research

NACE	Classification of Economic Activities in the European Community
NAICS	North American Industry Classification System
NCSES	National Center for Science and Engineering Statistics (US)
NEPAD	New Partnership for Africa's Development
Nesta	National Endowment for Science, Technology and the Arts (UK)
NESTI	OECD Working Party of National Experts on Science and Technology Indicators
NIS	National Innovation System
NPCA	NEPAD Planning and Co-ordinating Agency
NPI	Non-profit institution
NSF	National Science Foundation (US)
NPISH	Non-profit organisation serving households
NSO	National statistical office
OECD	Organisation for Economic Co-operation and Development
PSF	Policy Support Facility
R&D	Research and development
RIS	Regional Innovation Scoreboard
RICYT	Iberoamerican Network of Science and Technology Indicators
RIO	Research and Innovation Observatory
STI	Science, technology and innovation
SciSIP	Science of Science and Innovation Policy
SDGs	Sustainable Development Goals
SME	Small and medium-sized enterprise
SNA	System of National Accounts
SoS:DCI	Science of Science: Discovery, Communication, Impact
SPRU	Science Policy Research Unit
SR&ED	Scientific Research and Experimental Development

STEM	Science, technology, engineering and mathematics
STI	Science, technology and innovation
STIP	Science, Technology and Innovation Policy
STISA	Science, Technology and Innovation Strategy for Africa
TUT	Tshwane University of Technology
UIS	UNESCO Institute for Statistics
UK	United Kingdom
UN	United Nations
UNESCO	United Nations Educational, Scientific and Cultural Organization
UNFCCC	United Nations Framework Convention on Climate Change
UNIDO	United Nations Industrial Development Organization
UNSD	United Nations Statistics Division
UNU-MERIT	United Nations University – Maastricht Economic and social Research and training centre on Innovation and Technology, the Netherlands
US	United States
WIPO	World Intellectual Property Organization
ZEW	Centre for European Economic Research (Germany)

PART I

Introduction

1. Introduction to 'Innovation Everywhere'

1.1   A CHANGING WORLD – CAN WE MEASURE WHAT IS HAPPENING?

This book is about the measurement of innovation and the use of the resulting indicators to shape policy. The question that recurs throughout the book is whether, in a rapidly changing world, innovation can be measured everywhere, not just in the business sector.

This question is well established in the literature, but the world in which innovation happens is changing, as are the options for measurement and the ways of informing policy development. This suggests that measurement, and the means of influencing policy, be examined closely, taking note of the impact of the digital economy which is transforming business, governance and social activity.

Only recently have people had access to 'smart phones' that connect them to apps that help them travel, provide information when needed, tell them the state of the weather, stock market activity, or the trading value of their currency. If they want to assemble a lawnmower, a video can be found that takes even the most inept person through the process in easily understood steps. The phrase, 'there must be an app for that', whatever 'that' may be, is frequently confirmed. There is an app for that.

People are well served in the digital world by having their own email, iCloud storage, access to platforms where trades can be made, and knowledge exchanged. There are diagnostic apps that will interview patients and refer them to the right medical department and these machines learn. This is different from the 1990s, and earlier, when every application was written by a (human) programmer. Now, the app can learn from its interactions and revise its own algorithm. The digital divide (access to computers) is vanishing as smart phones and other digital devices spread and the knowledge divide (there is a computer available but there is no knowledge of what to do with it) is being removed, not by people but by machines that tell you what you can do with it. This raises another question of what you should or should not do with it.

In the 2020s there are new issues. The app will help you do what you want, but it, in return, wants to know where you are, to have access to your contacts and it may read your email and keep track of the web sites you visit. The justification is that the app can give better service to you if it knows more about you. However, there are other less friendly apps in cyberspace that want the same information to commit cybercrime of some kind. The world is digital and there are good things as well as bad things to be considered.

While the digital economy is changing the way in which people behave, it also is changing the way business and public institutions are functioning. Facial recognition is an issue in public space and artificial intelligence (AI), used by a business in its transactions with government, other businesses and people, is raising questions about privacy, confidentiality and security of the data being used.

The challenge is to produce indicators of innovation, to use the indicators to show what the outcomes of innovation policy are and then how to evaluate the policy so that it can meet the desired target. Measurement matters.

1.2   INNOVATION MEASUREMENT

Innovation is everywhere. New or improved products and processes are made available to potential users or brought into use by the business, institutions of higher education or households, but nowhere in the international standard definitions of innovation is there a statement of whether the innovation is good or bad. Financial services can produce product innovations that damage the economy, local government can move the disadvantaged from their communities to high-rise apartment blocks as an outcome of social innovation, and households, or individuals, can develop or modify products that improve their satisfaction, but may also increase risk. If desirable outcomes (inclusiveness, sustainability, jobs and growth …) are required, restrictions must be imposed on the measurement of innovation which may require more than one measurement over time to confirm the progress, or not, towards the desired outcomes.

What is different in 2020 is that the *Oslo Manual*, which provides guidelines for the measurement of innovation, has, in the fourth edition (OECD/Eurostat 2018), provided a general definition of innovation that can be used to measure innovation and innovation activities in the same way in all economic sectors. It also supports the measurement of the network that connects the actors in an innovation system. These linkages can be feedback loops that interact with one another, as well as with the actors, and result in non-linearity of response to policy interventions. This makes it difficult to predict the outcomes of changing policy. Innovation happens in a multi-connected complex system.

The fourth edition of the *Oslo Manual* has moved from being a manual used by European Union (EU) Member States and OECD Member Countries, and observers, to an international standard applicable in all countries. That is a significant step which is discussed in Chapter 6.

1.3   POLICY

Innovation policy is designed to promote innovation and the view over many years is that innovation occurs in the business sector and leads to jobs and growth. There is also a view that promoting research and development (R&D) gives rise to more innovation and in some countries innovation policy is R&D policy and R&D policy is tax policy. Are these the innovation policies for the 2020s and the digital economy? Changes in policy making in a digital economy, and how such policies are monitored and evaluated, are discussed in Chapter 10.

Innovation does occur in all economic sectors and it can support sustainable development, greater inclusion, and green outcomes to limit climate change. The question that this book addresses is how the implementation of these restrictions on innovation, and the outcomes, can be measured. Without measurement there is no monitoring of implemented innovation policy and no basis for evaluation which is needed for policy learning.

All countries, but especially developing countries, have innovation in their informal economy. Understanding how innovation happens in the informal economy is another researched field, but an important one, as it leads to the question of how innovation policy should influence innovation in the informal economy.

'Social innovation' is pervasive, but there is no internationally supported definition of social innovation. That does not make it any less important, but the discussion in Chapter 10 is whether social innovation can be measured and what a social innovation policy might look like.

1.4   STRUCTURE OF THE BOOK

The book has four parts. Part I comprises this introduction and a review of innovation systems in Chapter 2. Innovation systems occur in Part I as innovation is a systems phenomenon and innovation systems provide the framework for the topics in the rest of the book.

Part II starts with current innovation policies in Chapter 3, followed by scoreboards and their use for monitoring existing innovation policy in Chapter 4. Chapter 5 returns to innovation policy, but with a focus on implementation. The discussion is about whether innovation policy should involve the whole of government, or several separate policies managed by different government departments. Implicit is the view that innovation occurs in the business sector. Part II also raises questions about how to deal with innovation in all the economic sectors which are discussed in Part III.

Part III deals with the statistical measurement of innovation. Chapter 6 introduces the general definition of innovation applicable in all economic sectors, not just the business sector. This is a major step as it provides a standard definition at a time when the economy, and measurement of innovation in the economy, are changing rapidly. The chapter also introduces restricted innovation and how measurement takes account of restriction. Examples of restrictions already mentioned are 'inclusive' and 'sustainable' innovation. Innovation policy makers want to know about inclusion and sustainability as well as about jobs and growth. Chapter 6 also introduces the importance of language and how it is used in the innovation discourse.

Chapter 7 presents the conceptual framework for the specification of the statistical measurement of innovation. It is the application of the general definition of innovation (Chapter 6) to all economic sectors of the System of National Accounts (SNA), combined with the systems approach to innovation (Chapter 2). This provides guidance to measurement in developed and developing countries, an important change in the fourth edition of the *Oslo Manual* compared with previous editions. Another consequence of the use of the general definition of innovation is its application in the business sector. This was not done in the fourth edition of the *Oslo Manual* as the manual has been, and still is, a guide to measuring innovation in the business sector. Applying the general definition to the business sector and examining the role of zero price digital products raises questions addressed in Chapter 10.

While Chapter 7 deals with innovation in all economic sectors of the SNA, there are some forms of innovation that are not sector specific. Chapter 8 deals with them. They include innovation in the informal economy, eco-innovation for the green economy, social innovation and innovation resulting from the use of general purpose technologies and practices leading to the 'fourth industrial revolution'.

Part IV continues the discussion of the current state of innovation policy in Part II, and of statistical measurement of innovation in Part III, to pose the question, 'Where next?' Part IV starts, in Chapter 9, with a discussion of innovation and global challenges. The first is innovation and sustainable development and the 17 Sustainable Development Goals (SDGs) which are to be realised by 2030. The imposition on innovation of it being sustainable is an example of 'restricted innovation' raised in Chapter 6. The SDGs are followed by discussion of innovation and climate change and the green economy. Both are cases of restricted innovation. Chapter 10 considers the future of innovation in the digital economy, the informal economy and social innovation. It then looks at a series of issues that should stimulate innovation studies in the 2020s. These include the digital economy and its impact on innovation measurement, the Science of Innovation Policy and its impact on policy making and implementation. Chapter 11 concludes and poses some questions to the reader about where innovation measurement and policy are going.

2.   Innovation systems

2.1   INTRODUCTION

This chapter starts with a basic review of systems[1] as a means of classification of actors in the system, their activities and their linkages with
other actors. Activities can change with time. Linkages, and what they
transfer, can go in one or two directions and can also take time to make
the transfer. Linkages may act as feedback loops and they can interact
with one another, as well as with the actors, to alter the change antici-
pated when an activity is implemented. The system is bounded, which
means that there are boundary or framework conditions which influence
the activities of the actors and what flows through the linkages. A system
can be complex.[2]

After a general description of systems, 'innovation' and 'innovation
policy' are introduced, also at a basic level. Later, a connection is made
to national, regional and sectoral innovation systems.

The reason for this approach is to encourage the reader to reflect first
on what systems are and how they can be used to describe complex
phenomena. That understanding is a first step towards asking what could
be done to cause the system to behave differently and to learn from that.

Language is an issue throughout the book and the term 'eco-systems'
is an example. With one exception, 'systems' will not have the preface
'eco'. The exception is found in Chapter 9 when green innovation is
discussed, and ecological systems are relevant.[3]

2.2   SYSTEM COMPONENTS

In this book, the focus is on innovation everywhere in the economy.
As a consequence, the actors are 'institutional units', as defined in the
System of National Accounts Manual, 2008 (EC et al. 2009: 4.24).[4] These
include firms in the business sector,[5] government departments in the
general government sector, households, or individuals in the household
sector, and lobby groups, political parties, or other such groups, in the

non-profit institutions serving households (NPISH) sector. The public sector is the general government sector plus public institutions, such as government-controlled businesses, education institutions or hospitals. The SNA sectors are reviewed in Chapter 7.

The institutional units engage in activities, interact with other institutional units and produce outcomes in the short term that may lead to economic or social impacts in the longer term.

A point relevant to systems analysis, and especially to innovation systems, is that the actors do not act in isolation. It is the connection (linkage) with other actors that constitutes the system and allows some actors to influence the behaviour of others. This influence can result from new products made available to other actors (smart phones) or the response of regulators (actors) to services provided by other actors (e.g., germline gene editing). The term 'linkage' here covers any exchange, in any direction, between institutional units. Some examples are money, skilled people, property ownership, knowledge or services.

A system can be viewed as a set of actors engaged in activities, linked in various ways and producing outcomes, bounded by what have been called 'institutions' or 'rules of the game' (North 1990) or framework conditions, the term used here.

2.3   INTERVENTION, COMPLEXITY AND NON-LINEARITY

In an economic system, there are at least two kinds of interventions. The first acts directly on an actor to change its behaviour. For example, a government procurement office asks for tenders that require innovation to be part of the response. The behaviour of the successful bidder(s) changes as a result. The example used is the promotion of innovation in a firm. The change in behaviour could also include training of staff in the use of critical technologies (artificial intelligence), developing digital competence or meeting government requirements for an inclusive labour force.

The second type of intervention deals with the framework conditions which influence what the actors do. They could include policy to promote research in higher education, resulting in more highly qualified people who could be employed in the business sector or in public sector research establishments. Such a policy can influence innovation but not directly.

Policy intervention can be a mix of direct intervention and indirect intervention and this will be discussed in Part II.

Innovation systems are complex, non-linear and are difficult, if not impossible, to model. This is largely because of human behaviour which changes in response to interventions, direct or indirect, but not necessarily in the way expected. Non-linearity happens when interventions interact with one another or with links between actors. As a result, the desired output, more innovation perhaps, may not change in proportion to the intervention.

2.4 INNOVATION SYSTEMS

So far, systems have been introduced, but not innovation systems. That requires definitions of innovation and of innovation activities which may or may not lead to innovation. Once there is a definition of innovation (Chapter 6) the distinction can be made between institutional units that are 'innovative' and those that are not. Further distinctions can be made if the innovation is restricted (inclusive or sustainable innovation are examples).

The study of innovation, and innovation systems, has many approaches and one of the objectives of this book is to suggest a direction which can bring some, if not all, of these together.

One approach has been the measurement of innovation, and related activities, through internationally agreed surveys such as the EU Community Innovation Survey (CIS), using internationally agreed definitions. The evolution of the definition of innovation in four editions of the *Oslo Manual* is reviewed in Chapter 6, as well as the application of the definition in the series Community Innovation Surveys, starting with reference year 1992 and continuing to the present. The *Oslo Manual* provides definitions, the CIS provides data and the data are used to populate statistics which may be used as they are, or in combination with other statistics, as indicators. The principal use is the monitoring of the innovation system – the rate of innovation in the sector being studied, the allocation of resources to activities that support innovation and, perhaps, evidence of activities that support or inhibit innovation. These indicators may be used in 'scoreboards' which report on the innovation system in countries, supporting comparisons that may be used to influence policy (Chapter 4).

The indicators from the measurement approach to observing innovation can be used to monitor and to evaluate (Chapter 4) implemented innovation policy,[6] which may lead to policy learning and the further development of policy (Chapter 5). Part II reviews some current innovation policies, looks at the role of scoreboards in monitoring and evaluation, and at developing new policies or improving existing ones. With exceptions that will be noted, existing innovation policy is focused on the business sector. The discussion in Part II leads to the broader discussion in Part III on measuring innovation in other economic sectors.

The measurement approach to understanding innovation is used in statistical offices and in research institutes with a mandate to produce official statistics. The indicators resulting from the measurement are published by countries that produce reports on the state of their innovation system and by international and supranational organisations that produce scoreboards (Chapter 4). The definitions and guidelines for measurement, in the case of innovation, are developed and agreed by the OECD Working Party of National Experts on Science and Technology Indicators (NESTI) in collaboration with the EU statistical office, Eurostat. The membership of NESTI is a mix of statisticians who produce the innovation statistics and people from government policy departments of participating countries which use them. Staff from research institutes and academics are, in some cases, invited to join the country delegation to provide expertise on a specific topic.

Outside of the OECD/Eurostat community are academics who work on understanding innovation with a view to finding a theory of innovation. Some of this work is reviewed in the next section.

2.5   INNOVATION SYSTEMS LITERATURE

There is a large literature on innovation systems and on national innovation systems. In this section, both are considered briefly. The objective is to provide a background to the discussion of current innovation policy in Part II leading to questions to be considered in Part III when innovation in any economic sector is discussed.

An ongoing discussion is the theoretical basis for innovation policy. Borrás and Edquist (2019: 5) assert that 'the theoretical foundations for innovation policy design from the innovation systems approach remain underdeveloped today'. This is compatible with the approach of this book which focuses on statistical measurement guided by internationally agreed definitions indicating the outcome of implemented innovation policy.

Table 2.1   Key activities in innovation systems in Borrás and Edquist (2019: 25)

1. Provision of knowledge inputs into the innovation process
1.1 Provision of R&D results
1.2 Competence building
2. Demand-side activities
2.1 Formation of new product markets
2.2 Articulation of new product quality requirements
3. Provision of constituents for innovation systems
3.1 Creation and change of organisations
3.2 Interactive learning, networking and knowledge integration
3.3 Creation and change of institutions
4. Support services for innovating firms
4.1 Financing of innovation processes
4.2 Incubating activities
4.3 Provision of consultancy services

Source: An abbreviated version of Box 2.2 in Borrás and Edquist (2019: 25).

2.5.1   Innovation Systems

Edquist (2005) presents a chapter in the *Oxford Handbook of Innovation* (Fagerberg et al. 2005) which is an accessible discussion of innovation systems, their components and their purpose, which has had considerable influence over the last 15 years on innovation policy. It discusses functions or activities that are important for the analysis of innovation systems. This leads to a list of ten activities, or functions, that are important to the innovation system. The chapter reflects earlier work (Edquist 1997; Edquist and Hommen 2008). The most recent version of the list appears in Borrás and Edquist (2019: 25). It has the same content as that in Edquist (2005), but the activities are grouped under explanatory headings as shown in Table 2.1. These will be discussed in Chapter 6 when the definition of innovation and of innovation activities for measurement purposes (OECD/Eurostat 2018) are introduced.

For comparison, innovation activities were presented in the third edition of the *Oslo Manual* (OECD/Eurostat 2005: para. 40–43). In both cases, innovation, and innovation activities, took place in the business sector.

2.5.2   National Innovation Systems

Much of the literature in the last 30 years has addressed national innovation systems and Chaminade et al. (2018: 49) have provided a comprehensive review of the subject, noting that 'analysing the system is
very complex and there is a tendency to reduce complexity by looking at
particular aspects of the system …'. The key point is that an innovation
system is very complex which is a recurring theme in this book.

2.5.3   The Digital Economy and Innovation

The digital economy is growing rapidly and changing the way innovation
happens. This does not affect the innovation system, but it does influence
the speed at which it can happen and the means of providing digital
product innovations and digital process innovations. Some of this is discussed in Gault (2019) and in Paunov and Planes-Satorra (2019).

2.6   CONCLUSION

Innovation systems are complex and that is why there is no headline
indicator for the activity of innovation comparable to the ratio of gross
domestic expenditure on R&D (GERD) to gross domestic product
(GDP). To analyse what is going on in an innovation system may require
simplification of its description (Chaminade et al. 2018). It could also be
modelled, which is considered in Chapter 10.

   In the following chapters, the systems approach is used to describe
the innovation system, its components and its outcomes. In Part II, the
chapters deal with innovation in the business sector. Part III introduces
the general definition of innovation and the definition of the economic
sectors used by the SNA so that sector-specific innovation can be discussed. Chapter 8 considers types of innovation that can occur in any
sector. Part IV looks at what happens next, in this subject, in the 2020s.

NOTES

1.   See Meadows and Wright (2008) for more on systems from a basic
     perspective.
2.   See Mitchell (2009) for more on complexity.
3.   The view is that, in most cases, 'eco-' adds nothing to the discussion of
     innovation systems. For further discussion, see Borrás and Edquist (2019:
     26), Chaminade et al. (2018: 95) and Gault (2010: 32).

4. This is developed in Chapter 7.
5. In Chapter 8 the non-financial corporations sector and the financial corporations sector are combined to be the business sector. This aligns this book with the usage in the *Oslo Manual*.
6. The use of 'implemented' is important. Policies can be announced in various ways: the manifesto of a political party; the news media; or in the legislature. There is some distance between an announced policy and an implemented policy which can be monitored and evaluated.

PART II

Innovation policy

3. Innovation policy prior to 2020

3.1 INTRODUCTION

Innovation is credited with many things, including the creation of jobs and economic growth. Governments promote innovation in order to benefit from these outcomes and for decades, the emphasis has been on innovation in the business sector. That continues, but there is also interest in innovation in the public sector, and in types of innovation that can occur in any economic sector, such as innovation in the informal economy (Kraemer-Mbula and Wunsch-Vincent 2016) and social innovation (Moulaert et al. 2013; Moulaert and MacCallum 2019).

3.1.1 Some History

The study of innovation, and related policy, goes back to the 1930s (Schumpeter 1934), but it was not until 1992 that guidelines for the statistical measurement of innovation were formalised (OECD 1992). The European Union Community Innovation Survey (CIS) began, for reference year 1992, and provided the data to populate statistics used to produce indicators that could support the development of innovation policy. It is this connection between innovation policy and measurement that is the subject of this book.

Another high-level intervention in the development of innovation research, over the last 15 years, was a proposal by John Marburger, then the Director of the Office of Science and Technology Policy in the Executive Office of the President of the United States, to promote the 'Science of Science Policy'(Marburger 2005). Later in 2006, at the OECD Blue Sky Forum II (Marburger 2007), the proposal was extended to include innovation.

The National Science Foundation (NSF) responded by creating a programme for the 'Science of Science and Innovation Policy (SciSIP).[1] The NSF announced several SciSIP solicitations with a view to developing a community of practice, working on a better understanding of science and innovation policy and measurement, and a peer group of policy developers, managers and users.

The EU, the OECD and countries, including the US, were studying the need for science and innovation policy which, in turn, required statistical measurement and indicators. These issues are discussed further in Chapter 10, including the role of institutions doing research on science and innovation policy.

'Innovation', in the context of the SciSIP project, was that of the third edition of the *Oslo Manual* (OECD/Eurostat 2005). Products were new or improved and introduced on the market and processes were also new or improved and brought into use by the firm. If there was an implemented policy, its objective was to promote the activity of innovation in the business sector. The SciSIP project was to understand how innovation happened and how innovation policy could be improved. It was not there to propose policy but to examine which implemented innovation policies worked better than others.

A key point made by Marburger at the OECD Blue Sky Forum II (OECD 2007) was that 'in the face of a rapid global change, old correlations do not have predictive value' (Marburger 2007: 32). It was supported by Chris Freeman and Luc Soete, at the same forum (Freeman and Soete 2007: 272), who observed that 'the link between the measurement of national STI activities and their national economic impact, while always subject to debate, particularly within the context of small countries, has now become so loose that national STI indicators are in danger of no longer providing relevant economic policy insights'. The purpose of the OECD Blue Sky Forums was to examine the use of current indicators of science, technology and innovation (STI) and the need for better indicators and new indicators. Both Marburger (2007) and Freeman and Soete (2007) provided serious challenges to make STI indicators more relevant and SciSIP was one approach to a solution for such problems (Gault 2011).

The SciSIP work has gone on, supported by NSF, and Al Teich (2018) has provided a comprehensive review of the SciSIP process and its achievements in more than a decade of research. Creating a new specialty in the social science community takes time, and that goal is yet to be realised.

Meanwhile, in September 2019, a 'Dear Colleague Letter' was issued by NSF (2019) which made the following statement about SciSIP:

> Science of Science: Discovery, Communication, and Impact
> https://www.nsf.gov/funding/pgm_summ.jsp?pims_id=505730: This evolution of the former Science and Innovation Policy Program focuses on basic research that can increase the productivity of scientific workflows, our nation's capacity to communicate it accurately and effectively, and the value of that work to society.

The Science of Science: Discovery, Communication, Impact (SoS: DCI) Program removes 'innovation' and 'policy' to focus on the science of science. Fortunately, there are many research groups studying innovation policy. Another Marburger (2007) proposal is discussed in Chapter 10 and his work is reviewed in Marburger and Crease (2015).

3.1.2   International and Supranational Organisations

Because of the importance of innovation, international and supranational organisations publish information on innovation policy (OECD 2011, 2014;[2] EU 2013; GTIPA 2019) and delegations discuss policy issues in their committees. The EU provides fact sheets to guide researchers, one of which is The Innovation Policy Fact Sheet.[3] No one expects there to be a single solution to implementing effective innovation policies, but the objective is to see what works and what does not in other countries. This has given rise to an extensive literature on the theory and practice of innovation policy (Smits et al. 2010; Edler et al. 2016) and research institutes that focus on the subject (the Science Policy Research Unit at the University of Sussex and the Manchester Institute of Innovation Research and its predecessors at the University of Manchester are examples in the UK. Internationally, there is the European Forum for the Studies of Policies of Research and Innovation (EUSPRI)).

3.1.3   OECD and Innovation Strategy

The OECD published an innovation strategy in 2015 (OECD 2015a), an update of the 2010 strategy (OECD 2010a) which had a separate report on measurement (OECD 2010b). The 2015 strategy provides advice, rather than specific policies for countries. It proposes five characteristics of a positive environment for innovation (OECD 2015a: 12):

- a skilled workforce
- a sound business environment
- a strong and efficient system for knowledge creation and diffusion
- policies that encourage innovation and entrepreneurial activity
- a strong focus on governance and implementation.

These characteristics are supported by five priorities (OECD 2015a: 12). They are to:

- strengthen investment in innovation and foster business dynamism
- invest in, and shape, an efficient system of knowledge creation and diffusion
- seize the benefits of the digital economy
- foster talent and skills and optimise their use
- improve the governance and implementation of policies for innovation.

The positive environment and the priorities demonstrate that countries have, to some extent, moved on from treating R&D policy as innovation policy, although countries with a small business sector may still focus on R&D rather than innovation.

The concept of a system was discussed in Chapter 2. The importance of the digital economy is a priority for the OECD innovation strategy, and this in 2015 when the digital economy was just beginning to have an impact. Finally, governance and implementation of polices are priorities. Both are discussed in Chapter 5.

The OECD innovation strategy (OECD 2015a) does not deal with sectors other than the business sector. Innovation in the public sector is considered in a separate publication (OECD 2015b). In the 2010 version of the OECD innovation strategy, the statistical measurement of innovation was treated separately (OECD 2010b). This is discussed in Part III.

3.1.4   European Union and Innovation Policy

The Innovation Policy Fact Sheet describes activities related to policy and provides links to innovation projects.

The EU innovation policy is not isolated. It is linked to policies on employment, competitiveness, environment, industry and energy and the role of innovation is explicit 'to turn research results into new and better services and products in order to remain competitive in the global marketplace and improve the quality of life of Europe's citizens'. The linking of policy objectives is consistent with a systems approach to innovation policy.

In 2010, the EU introduced the Innovation Union and it has produced the Innovation Union Scoreboard (discussed in Chapter 4), the Regional Innovation Scoreboard and the Innobarometer. The Innobarometer conducts an annual opinion poll asking businesses, and the public about innovation policy. The statement that 'innovation is made possible by research and education' makes clear the role of formal knowledge generation and of institutions of education.

A key point is that Horizon 2020, the EU's 8th Framework Programme (2014–20), is the first framework programme to integrate research and innovation. It also supports work on public sector and social innovation.

In 2015, the Commissioner responsible for research, science and innovation announced the European Innovation Council,[4] which is under consideration for the successor to Horizon 2020. It is managed by members of the business community and it provides advice and support to businesses.

While the promotion of innovation through innovation policies is an active and current topic, there is also interest in how knowledge moves from one unit to another and how to support the transfer of knowledge from research institutions to firms that could then use it to support innovation (EU 2014).

In summary, there is significant interest in the EU and support for innovation in the business and the public sectors and for social innovation. The knowledge gained from this involvement could contribute to manuals, like the *Oslo Manual*, on the measurement of innovation in sectors other than the business sector and for innovation, in this case social innovation, that occurs in any economic sector. There is also interest in open innovation and the flow of knowledge between institutional units.

3.2   INNOVATION IN STATISTICAL SURVEYS

After years of R&D policy dominating innovation policy, there is ample empirical evidence to show that, in the business sector, more firms innovate than do R&D. Table 3.1 provides information from the NSF 2014

Table 3.1   *More firms innovate than do R&D*

Firm characteristics	All firms	Product or process innovations	No. of innovative firms
	Number (000)	Per cent	Number (000)
All	1273.3	15.4	196.1
With R&D	53.5	69.5	37.4
No R&D	1219.9	13.1	159.8

Note: That more firms innovate than do R&D is not peculiar to the US. Any country that runs an innovation survey and asks questions about R&D should be able to come to the same conclusion.
Source: Kindlon and Jankowski (2017: Table 3) and author calculation.

Business R&D and Innovation Survey (BRDIS) (Kindlon and Jankowski 2017).

Two points in Table 3.1 are important for innovation policy. The first is that the propensity to innovate is higher for firms that do R&D (69.5 per cent) than those that do not (13.1 per cent). The second is that 159,800 firms innovate but do no R&D compared with the 37,400 firms that innovate and do R&D.

There are two other empirical observations that should influence policy development. They are the correlation between the propensity to innovate and the size of firms, measured by employment (Kindlon and Jankowski 2017: Table 2), and the correlation between the propensity to innovate and the expenditure on R&D by the firm (Kindlon and Jankowski 2017: Table 3). A summary of these observations is that large firms are likely to perform R&D and to innovate, but there are many more small firms than there are large firms. What are the policies for them, and do they take account of firm size?

The innovation policy question is illustrated in Table 3.2 which can be populated with data from any country that runs an innovation survey. From Table 3.1, the number of firms in C is larger than in E and the policy question is whether, or not, to promote the performance of R&D by firms in C. Before doing that, the size effect should be examined, producing Table 3.2 for a series of firm sizes. There are firms that engage in innovation activities, but do not innovate, and some will do R&D (D) and some not (B). The policy question is whether to provide an incentive to firms in B to move to C and firms in D to move to E. Firms could also be encouraged to move from B to D and then to E. What remains is how to deal with firms in A.

Table 3.2   *Reporting innovation in the business sector*

	Innovative Firms	Non-Innovative Firms		Total Firms
		Ongoing and/ or abandoned innovation activities ONLY	No innovation activities	
R&D performed	E	D	n/a	
R&D not performed	C	B	A	
Total Firms				

Note: This table was first presented at a workshop of the African Science, Technology and Innovation Indicator (ASTII) initiative in Windhoek, Namibia on 25 May 2017.
Source: The author.

A policy consideration is whether to promote the growth of firms, assuming that as the firm grows in size, the firm strategy will recognise the need for the performance of R&D if it is going to be able to bring product innovations to the market. Promoting growth is also an issue in countries that provide R&D incentives to small firms, with no incentive or a less attractive incentive to larger firms. Such a policy may promote R&D, but not growth.

3.3   SOME INNOVATION POLICIES

Innovation policies over the last five years have had many objectives. Some are reviewed here.

3.3.1   Innovation Related to People

Innovation requires knowledgeable people, and this leads to support for institutions of higher education to produce more graduates in the science, technology, engineering and mathematics (STEM) fields. This objective also has implications for primary and secondary school education and the need to prepare students for the courses they will enter in post-secondary institutions. It also raises a question of content.

With the use of artificial intelligence (AI) as part of innovation, there is a need to have people involved in innovation activities with a good knowledge of sociology and psychology as the new or improved products being offered, as product innovation, are designed to change the behaviour of groups, and of individuals. This does not preclude the need for people with technical skills.

In addition to education, the health of people matters, if they are to be productive, and health is linked to human behaviour. After years of consuming food containing animal fat, salt and sugar, there is a population of obese people more likely to have type 2 diabetes, cardiovascular illness and the need to have hip and knee joints replaced. This is an example of a systems problem where the policy has to address the cause of the problem while also supporting medical research to deal with the immediate need to care for people with medical problems. Time is an important variable in the development of innovation policy and the analysis of the implemented policy.

Demographic distributions can influence innovation policy. In the developed world, there is an ageing population and people need new ways to support them in their old age. While this is a humanitarian undertaking, learning how to make the lives of the elderly better can lead to marketable products (Gault 2010: 106; Aho et al. 2013). In many developing countries, especially in Africa, the median age is quite low (in Rwanda it is 19) and the challenge is to engage a young population in productive activities.

3.3.2   Knowledge Generation and Transfer

Knowledge generation
The formal generation of knowledge, R&D, is an innovation activity (OECD/Eurostat 2018: 85) and, as with all other innovation activities, it does not necessarily result in innovation. R&D can happen in all sectors of the economy and there are innovation policies directed at promoting R&D in the business sector and in the public sector. There is also policy in support of transferring the knowledge from research institutions and institutions of higher education to firms in the business sector, either directly or through intermediaries.

The OECD has reviewed R&D tax support and introduced a new OECD R&D Tax Incentives Database (Appelt et al. 2019). One of the OECD countries that does not have R&D tax incentives is Germany, but one is under review following the tabling by the government of a bill in May 2019. The bill, if it becomes law, will permit 25 per cent of eligible expenses on R&D (calculated on the basis of the cost of R&D personnel), up to €2,000,000 per year, to be claimed as tax credit (the credit has a limit of €15,000,000 per R&D project). This also applies to R&D

that has been contracted out. In this case the rate (of eligible expenses) remains at 25 per cent, but it applies only to 60 per cent of the personnel costs.

The EFI 2019 report (EFI-Commission of Experts for Research and Innovation 2019) supports an R&D tax credit and refers to the 2017 EFI report (EFI-Commission of Experts for Research and Innovation (2017: 9) for potential courses of action. There are two: a tax credit on income tax proportional to a company's internal R&D expenditures, or a tax credit on wage tax calculated on the basis of the R&D personnel costs incurred. The second was recommended. In both cases, the tax incentive is aimed at small and medium-sized enterprises (SMEs). However, the draft bill does not limit the potential tax credit to SMEs. Any firm may apply.[5]

Focusing on SMEs, depending upon the type of R&D tax credit, can create a barrier beyond which a SME may not wish to grow if it loses the tax incentive or the tax incentive changes. This issue is discussed by Drummond and Bentley (2010) in relation to the Scientific Research and Experimental Development (SR&ED[6]) tax benefit programme of Canada.

Another observation about R&D is that the population of R&D performers in the business sector is dominated by R&D performing firms that do little R&D and are sporadic in their performance. This has been demonstrated by Molotja et al. (2019) for South Africa and it confirms work done in Canada (Schellings and Gault 2002). See also Rammer and Schubert (2016).

Knowledge transfer
There are a number of policies promoting the transfer of knowledge in support of innovation. Firms can acquire R&D results from outside of the firm if they have the capacity to absorb the knowledge. Policies dealing with education and training, which may be government policies or firm strategy, or both, can support this transfer of formal knowledge and its application.

The application of firms for vouchers which, if granted, allow the firm to pay institutions of higher education, or research institutes, for help in solving problems is becoming more common (OECD 2014: 160). It is also a way for firms that do no R&D to gain knowledge needed for innovation.

Knowledge transfer is not limited to technical knowledge. Innovation policy can include the financial support for start-ups and the provision of knowledge needed to manage the funding. In August 2019, the German Federal Ministry of Economic Affairs and Energy released a list of funding programmes (BMWi 2019).[7] The programmes are designed to support different financial needs at different stages of development of the client firms. One, the High-Tech Start-up Fund (HTGF), provides early-phase funding but also 'ensures that the management of young start-ups receives the necessary help and support'. This goes beyond financial support and supports a wider range of knowledge to be transferred.

While financing of innovation activities in firms is important, so also is proximity. Uyarra and Ramlogan (2016) have examined the effect of cluster policy on innovation. Their paper is part of the Compendium of Evidence on the Effectiveness of Innovation Policy Intervention Project undertaken by the Manchester Institute of Innovation Research (MIoIR), funded by the National Endowment for Science, Technology and the Arts (Nesta).[8] At the OECD, there is earlier work on clusters (OECD 2009) where case studies are presented in seven countries and policy issues are reviewed. Since 2012, the Cluster Observatory, run privately by CSC in Stockholm,[9] has supported researchers, policy makers and cluster organisations.

Dealing with the future
Country innovation policies address improvements for dealing with agriculture and food processing, renewable energy production and use, and transportation efficiency. The focus is more on what will happen than what is happening as a result of innovation in these areas.

Globally, there are the 17 UN Sustainable Development Goals (SDGs) and the objective of meeting the SDG targets by 2030. Eco-innovation is part of an international discourse and there is a plan to produce a manual on the subject (Kemp et al. 2019) to guide statistical measurement, analysis and policy learning. These are addressed in Chapter 9.

3.4   INNOVATION POLICY IN COUNTRIES

The OECD provided country profiles, including innovation policies, up to the 2014 Outlook (OECD 2014). The OECD-World Bank Innovation Policy Platform (IPP) has also been used as a valuable tool for understanding innovation policy, but it was archived on 1 July 2019. A source of country information is the European Commission/OECD International Survey on Science, Technology and Innovation Policy (STIP). The current version is the 2017 survey. The electronic platform is 'Research and Innovation' from the Research and Innovation Observatory – Horizon 2020 Policy Support Facility (RIO – H2020 PSF).[10]

While the emphasis in the RIO material is on R&D, the search facility provides access to 'vouchers' and to those countries using a voucher policy for innovation. The digital economy is present, but it draws on the IPP which is no longer supported and does not include the most recent OECD Digital Economy Outlook (OECD 2017e). However, the RIO – H2020 PSF is a potentially valuable tool that will grow in usefulness as it identifies and includes additional policy relevant topics.

The STIP Compass International Database on STI policies[11] provides access to the policies of participating countries. For example, a query on 'vouchers' generates 132 responses, 258 for the digital economy, 155 for social innovation. These are topics discussed later. The database also shows how many STI policy activities are engaged in by each country. This informs discussions, or recommendations, about 'whole of government', 'holistic' or 'fragmented' innovation policy.[12]

In summary, these two databases provide relevant information on the innovation policies of countries and can be used by the reader to pursue issues raised later in this book. While there are many references to public sector innovation, the work cited was before the provision of a general definition of innovation in all economic sectors. The general definition is introduced in Chapter 6.

3.5 NATIONAL INNOVATION SYSTEMS AND POLICY

So far, the discussion has been on innovation policy with reference to innovation systems (Chapter 2) and statistical measurement of innovation. The extensive literature on National Innovation Systems (Freeman 1987; Lundvall 1992; Nelson 1993) has not been considered. Part of the reason is that this book is focused on a definition of innovation for use in all economic sectors (Chapter 6), supporting statistical measurement of innovation and innovation activities (Chapters 7 and 8) which lead to the production of innovation indicators that can be used to develop innovation policy and then to monitor and evaluate innovation policy, once it is implemented, leading to policy learning. The National Innovation System (NIS) goes beyond this basic approach.

The definition of a NIS used by Lundvall et al. (2009) and also by Chaminade et al. (2018) is the following:

> ... an open, evolving and complex system that encompasses relationships within and between organisations, institutions and socio-economic structures which determine the rate and direction of innovation and competence-building from processes of science-based and experience-based learning.

This is a definition of a system, specifically a NIS, and it is not to be confused with a definition of innovation for the purpose of statistical measurement. It is elaborated upon in Chaminade et al. (2018: 70–3) and it is considered here because of its importance for innovation policy.

The emphasis is on learning and the diversity of learning suggesting that related policy deals with how learning is supported, whether it is experience-based or science-based. The different needs, and policies, in developing countries are noted. In Chapter 2 of this book, the system consisted of institutional units, engaged in activities, having linkages with other institutional units, resulting in outcomes and longer term impacts. Considering the NIS, learning could take place in the institutional unit as an innovation activity, or through a linkage which could be formal or informal with another institutional unit (a firm or a research institution, for example). All of these possibilities suggest different policy interventions.

The NIS definition stresses linkages not just with other institutional units but within the institutional unit. The linkages could include global value chains which may require a change in the organisation of the production of the institutional unit connected to the value chain.

Chaminade et al. (2018) raise the importance of the informal economy, especially in developing countries (Kraemer-Mbula and Wamae 2010; Kraemer-Mbula and Wunsch-Vincent 2016) and note the importance of understanding innovation, learning and competence building within the NIS. This topic is discussed further in Chapter 8, along with the implications for policy.

Structural issues in the NIS are raised, both economic and social, which could be regarded from a systems perspective as framework conditions as discussed in Chapter 2. These framework conditions may influence government policy, business strategy and behaviour of individuals and households.

3.6   CONCLUSION

This chapter has looked at innovation policies now and how they are approached by the EU and the OECD, as well as by selected countries. Innovation has been confined to the business sector which is seen as a source of jobs and economic growth. While there is some reference to the public sector, and to social innovation, there is no single agreed definition of innovation outside of the business sector. The definition is an issue for Chapter 6 and then the application of the definition takes place in Chapter 10.

Presented in this chapter are the different approaches to innovation policy and what drives it. The question remains as to what extent R&D, or knowledge gained from experience, dominate innovation and innovation policy.

NOTES

1.   This programme is reviewed in Teich (2018) and is further discussed in Chapter 10.
2.   OECD (2014) is cited, rather than more recent Outlooks, because it provides, in Chapter 9, STI country profiles which are still worth reading. Country profiles do not appear in the next two Outlooks, OECD (2016) and OECD (2018b), but note the change of the name of the publication with 'Innovation' replacing 'Industry'. In the Foreword to OECD (2016) reference is made to the European Commission/OECD International Survey on Science, Technology and Innovation Policy (STIP) as a source of indicators to monitor innovation policy. The OECD-World Bank Innovation Policy Platform (IPP) is also referenced. It was a valuable tool for understanding innovation policy, but it was archived on 1 July 2019. In the 2018 Outlook, use of the 2017 STIP (https://stip/oecd.org, accessed 17 March 2020) is acknowledged and it is a useful source of information on innovation policy.
3.   See   http://www.europarl.europa.eu/factsheets/en/sheet/67/innovation-policy (accessed 17 March 2020).
4.   See https://ec.europa.eu/research/eic/index.cfm (accessed 17 March 2020).
5.   This is the situation at the time of writing. Bills can be amended or withdrawn. If the tax credit is of interest, the reader is recommended to follow the progress of the bill. A source in English is Leyton.com.
6.   See https://www.canada.ca/en/revenue-agency/services/scientific-research-experimental-development-tax-incentive-program.html   (accessed   17 March 2020).
7.   See   https://www.bmwi.de/Redaktion/EN/Dossier/financing-for-start-ups-company-growth-and-innovations.html (accessed 17 March 2020).
8.   The   20   reports   are   found   at   http://www.innovation-policy.org.uk/compendium/ (accessed 17 March 2020).

9.   See http://www.clusterobservatory.eu/ (accessed 17 March 2020).
10.   See also Edler et al. (2016).
11.   https://stip.oecd.org/ (accessed 17 March 2020).
12.   See Borrás and Edquist (2019) for a view on 'holistic' innovation policy.

4. Monitoring and evaluation of innovation policy

4.1 INTRODUCTION

Chapter 3 discussed existing innovation policies and strategies, and their sources. It introduced statistical measurement that supports research on the subject of innovation, influenced by policy that has been implemented. This chapter looks at the monitoring and evaluation of implemented innovation policy and how this is done. Four topics are covered: country reports; reviews of innovation policy conducted by independent experts; scoreboards that can be used to rank countries; and country ranking.

Country reports of innovation surveys can be used to support research on innovation (discussed in Chapter 3) and to promote a wide range of discussion of the findings from innovation surveys conducted in the country. Such discussion in statistical offices, policy departments, the business sector, and other sectors, including civil society, can influence subsequent surveys, analysis and policy development. The real advantage of producing clear and accessible country reports, and distributing them widely, is the development of an informed community that can offer constructive criticism of implemented innovation policy.

Innovation policy reviews are conducted by experts at the request of the country that has developed and implemented the innovation policy. The most recent example is the OECD (2019b) review of higher education, research and innovation for Portugal. The importance of such reviews is that they bring together a community of policy makers, users of the policy and statisticians, who can benefit from interacting with the experts that conduct the review.

Scoreboards gather data on countries from a variety of sources and produce indicators related to innovation. Some of these indicators come from direct measurement of innovation by innovation surveys, others deal with framework conditions. Examples are provided from the OECD, the European Commission, the Global Innovation Index (GII) and the *African Innovation Outlook* (AIO). Scoreboards can raise the discussion of innovation from a country level (country reports) to an international level and lead to the adoption of best practices by participating countries. Scoreboards can also be used for ranking of countries.

Ranking of countries is quite different from monitoring of implemented innovation policy as the ranking can have impact on the behaviour of the countries being ranked which may not help evaluate and improve innovation policy.

The country reports, policy reviews by independent experts and scoreboards, all deal with innovation in the business sector, with one exception, the EU public sector innovation scoreboard. It is presented, along with a brief discussion of household innovation, as an introduction to monitoring innovation in all economic sectors. This is discussed in Chapter 7 while this chapter looks at what is being done now.

4.2   COUNTRY REPORTS

4.2.1   Some Examples

EU Member States, and Iceland and Norway, conduct the Community Innovation Survey (CIS) every two years and report their findings to Eurostat, the statistical office of the EU. They may also produce country reports, an example of which is the UK report for the CIS 2014–16 released in 2017.[1]

Outside of the EU, examples of country reports are found in Canada, which runs a survey on innovation and business strategy,[2] and the US *Science and Engineering Indicators 2018* (National Science Board 2018: Chapter 8).

An example of a report on a specific topic is the US *InfoBrief* (Kindlon and Jankowski 2017). The points were made in Chapter 3 (Table 3.1) that more firms innovate than do R&D, but firms that do invest in R&D have a higher propensity to innovate. A more comprehensive discussion of innovation indicators is provided in *Science and Engineering Indicators 2018* (National Science Board 2018: Chapter 8).

If R&D promotion is part of innovation policy, it is important to monitor the link between R&D expenditure and innovation. If innovation policy is directed at firms that do not do R&D, again, they should be monitored, along with firms that move to and from R&D performance (Molotja et al. 2019). The paper by Molotja et al. (2019) examines the presence of firms in a ten-year period. The firms are classified according to their level of expenditure on R&D performance and the number of years they are present in the ten-year period. There are two findings relevant to innovation policy. The first is that large spenders on R&D performance are relatively few but have a high likelihood of being present for all ten years. The second is that most of the population is accounted for by small performers of R&D that are present for one or two years in the ten-year period.

Innovation surveys and country reports are not just the prerogative of developed countries. In Africa, there have been two editions of the *African Innovation Outlook* (AU-NEPAD 2010; NPCA 2014) which have chapters on R&D and innovation activities, and policies, in participating countries. These are closer to country reviews than scoreboards as the innovation activities of countries are deliberately not ranked. A third report was released in 2019 (AUDA-NEPAD 2019). A specific example of a country report is provided by Kenya (Ministry of Education, Science and Technology, Kenya 2016).

4.3 COUNTRY REVIEWS

Country reviews of innovation policy are provided by the OECD at the request of countries,[3] the five most recent reviews are for Portugal (OECD 2019b), Austria (OECD 2018a), Kazakhstan (OECD 2017a), Norway (OECD 2017b) and Finland (OECD 2017c). Country reviews are also conducted by other international organisations such as the World Bank,[4] UNIDO[5] and UNESCO, through its Global Observatory of Science, Technology and Innovation Policy Instruments (GO-SPIN) programme.[6]

While the methods used in the reports vary, the emphasis is on policy rather than statistical measurement and on what the countries are doing rather than on international comparisons. Scoreboards, in the next section, provide country comparisons.

4.4 INTERNATIONAL SCOREBOARDS

4.4.1 Introduction

International scoreboards vary in the indicators that they report, but they all provide a means of ranking the countries studied. In principle, this allows countries to review what they are doing and to improve activities that result in a low ranking. Whatever the cause, the ranking could be regarded as a warning which needs action. However, countries could be driven by the ranking rather than assessing their innovation system, reviewing that and committing resources that better serve the country than allocating the resources to moving up the ranking of a scoreboard. There is also a question about the extent to which indicators are linked. Action on one might have unexpected consequences for others.

In what follows, three scoreboards are discussed: the OECD Science, Technology and Industry Scoreboard 2017 (OECD 2017d), the European Innovation Scoreboard (EIS) (European Commission 2019) and the Global Innovation Index (Cornell University, INSEAD and WIPO 2019). The innovation that is presented in these scoreboards takes place in the business sector and there is limited reporting of direct measurement of innovation, but more on the framework conditions, introduced in Chapter 2, for innovation to occur. There is also considerable coverage of R&D, which is an innovation activity but not innovation unless the result of the R&D is introduced on the market or brought into use by the firm (OECD/ Eurostat 2005). A comprehensive review of innovation scoreboards is found in Hollanders and Janz (2013).

While the examples in this section deal with business sector innovation, using the definitions of innovation in the third edition of the *Oslo Manual* (OECD/Eurostat 2005), there has been work done on innovation in other sectors, especially the public sector, discussed in Chapter 7 but without the advantage of a standard definition of innovation applicable in all economic sectors.

4.4.2   OECD Science, Technology and Industry Scoreboard 2017

The OECD 2017 Scoreboard highlights the digital economy and the technologies that are part of it. Specific discussions of innovation are found in Chapters 3 and 4 of the Scoreboard. The digitalisation of the economy is changing it and the ways in which innovation can happen. This is discussed further in Chapter 10 of this book.

Chapter 3 of the Scoreboard, 'Research Excellence and Collaboration', deals with collaboration leading to innovation. In most countries, collaboration is led by suppliers and clients of the firm that innovates while there is less collaboration with institutes of higher education or research institutions. There is a firm size effect as large firms have greater absorptive capacity and are more likely to collaborate with institutes of higher education or research institutions. SMEs are more likely to collaborate with suppliers and clients. The size effect also applies to collaboration with other organisations as part of work on innovation.

Chapter 4 of the Scoreboard, 'Innovation in Firms', provides a country comparison of innovative firms in information and communication technologies (ICT) manufacturing and information technology services for the reference period 2012–14 and notes that 74 per cent of ICT manufacturing firms innovated, compared with 51 per cent for all firms in manufacturing (OECD 2017d: 152). This has implications for analysing the digital economy while noting that the digital economy is much more than the ICT sector.

Data are provided to demonstrate the influence of firm size on innovation (OECD 2017d: 154) by type of innovation and by new to market product innovation. There is further analysis of size dependence of firms receiving public support for innovation (OECD 2017d: 159). The firm size dependence of innovation is an issue for the formulation of innovation policy of governments (Chapter 3).

The sources of the innovation data used in Chapters 3 and 4 are the OECD survey of national innovation statistics and the Eurostat Community Innovation Survey (CIS 2014).[7] The point to note is that the OECD Science, Technology and Industry Scoreboard 2017 makes extensive use of the direct measurement of innovation while providing comments on the measurability of innovation.

4.4.3 The European Innovation Scoreboard 2019 and the Regional Innovation Scoreboard 2019

The European Innovation Scoreboard (EIS) 2019 (European Commission 2019) presents information on EU Member States and selected countries outside of the EU. The Member States are classified into four performance groups, based on their average performance scores as calculated with a composite indicator.[8] There are four performance groups (European Commission 2019: 6):

1. **Innovation leaders**: Denmark, Finland, the Netherlands and Sweden
2. **Strong Innovators**: Austria, Belgium, Estonia, France, Germany, Ireland, Luxembourg and the United Kingdom
3. **Moderate Innovators**: Croatia, Cyprus, the Czech Republic, Greece, Hungary, Italy, Latvia, Lithuania, Malta, Poland, Portugal, Slovakia and Spain

4.   Modest Innovators: Bulgaria and Romania.

As with the OECD Science, Technology and Innovation Scoreboard 2017, many of the indicators are not directly related to innovation but related to possible inputs and outputs of innovation. Human resource information and publications and patents are examples.

Of the 27 indicators used in the Scoreboard, six are direct measures of innovation, or innovation activity, taken from the CIS. The six are described in Annex E of the EIS. Each indicator has a number and the CIS-based indicators, with their indicator numbers, are the following:

2.2.2 Non-R&D innovation expenditures (percentage of turnover)

3.1.1 SMEs introducing product or process[9] innovations (percentage of SMEs)

3.1.2 SMEs introducing marketing or organisational innovations (percentage of SMEs)

3.1.3 SMEs innovating in-house (percentage of SMEs)

3.2.1 Innovative SMEs collaborating with others (percentage of SMEs)

4.2.3 Sales of new to market and new to firm innovations as percentage of turnover.

Indicators 3.1.1, 3.1.2, 3.1.3 are direct measurements of innovation in SMEs. Indicator 3.2.1 captures the effect of networks and collaboration. Indicators 2.2.2 and 4.2.3 apply to firms of all sizes and deal with the expenditure on innovation activities that are inputs to innovation (2.2.2) and sales of product innovation that are outputs of innovation. Data on SMEs are relevant because SMEs form the majority of firms in most countries and can play a vital role in innovation, for example, as developers of new ideas and as adopters of new technologies (Hollanders and Janz 2013). SMEs also tend to be single establishment firms which avoids the multi-establishment issues of analysing the data for large firms.

Countries are ranked by findings in the Scoreboard and the ranking may influence policy development. This gives rise to challenges of the methodology (Edquist and Zabala-Iturriagagoitia 2018) and discussion in governments and civil society.

The Regional Innovation Scoreboard (RIS)[10] 2019 is an extension of the EIS which covers 239 regions in 23 EU countries, Norway, Serbia and Switzerland. It uses 18 of the 27 indicators in the CIS and the principal finding is that Europe's most innovative regions are in the most innovative countries.

4.4.4   Global Innovation Index (GII) 2019

The Global Innovation Index (Cornell University, INSEAD and WIPO 2019) has been in place since 2007 and provides country reviews and rankings that are highly influential. In addition to its annual indicator report, the GII addresses a theme, for 2019, *Creating Healthy Lives – The Future of Innovation*, and for 2018, *Energising the World with Innovation*. It is in the theme chapters that possible innovations are presented.

The 80 GII indicators deal with framework conditions for innovation and are grouped into seven themes.

1.   Institutions
2.   Human capital and research
3   Infrastructure
4   Market sophistication
5.   Business sophistication
6.   Knowledge and technology outputs
7.   Creative outputs.

Each theme has two subsections where the indicators are described, and their sources noted. While the statistical measurement of innovation is not discussed, there is enough information on topics related to innovation to support the monitoring and evaluation of innovation policy.

4.4.5   Scoreboard Summary

This section has presented three types of scoreboard: the OECD Science, Technology and Industry Scoreboard 2017 which uses and presents statistical measures of innovation from a number of sources, the EU Innovation Scoreboard 2019 which makes use of CIS results in the six innovation indicators (of 27) focused on innovation in SMEs, the Global Innovation Index 2019 (Cornell University, INSEAD and WIPO 2019) which does not address statistical measurement of innovation but provides 80 indicators of framework conditions which could support or impede innovation. Examples of innovations are provided in the chapters on the theme of the year, *Creating Healthy Lives – The Future of Innovation*, in the case of the 2019 GII.

There are other scoreboards but these three provide examples of different approaches to innovation indicators derived from measures of innovation that take place in the business sector.

4.5   RANKING

Ranking is an outcome of scoreboards and it can have negative impacts. A country that is low in the ranking may have difficulty attracting a skilled labour force. As already discussed, countries may allocate resources to try to improve their ranking, to the detriment of their economy and society.

If countries are going to respond to scoreboards and their rankings, they should convince themselves of the legitimacy of the scoreboards and their contribution to society (Davis et al. 2012). This is not an issue for the scoreboards reviewed in this chapter, but it is an important consideration as economies change and new scoreboards appear. Von Bogdandy and Goldmann (2012) provide an illustration of the power of scoreboards.

4.6   INNOVATION BEYOND THE BUSINESS SECTOR

For the measurement of innovation in the business sector, there was the definition of innovation from the third edition of the *Oslo Manual* (OECD/Eurostat 2005). There was no comparable definition for innovation in the general government sector or the public sector (general government and public institutions), an issue discussed in Gault (2018a), and resolved in the fourth edition of the *Oslo Manual* (OECD/Eurostat 2018).

The absence of an internationally accepted definition of innovation did not prevent work on understanding innovation in the public sector.[11] Some of the projects are discussed in Chapter 7 of this book, including the European Public Sector Innovation Scoreboard 2013 (EPSIS)[12] (EC 2014) and the OECD Observatory of Public Sector Innovation (OPSI). The *Handbook of Innovation in Public Services* (Osborne and Brown 2013) provides a comprehensive review of innovation in the public sector. The general definition of innovation (OECD/Eurostat 2018), applicable in the public sector, is also discussed in Chapter 7.

From a historical perspective, it is instructive to see the evolution of the definition of innovation for use outside the business sector going back at least a decade. In the Nordic countries, there was the 'Measuring Public Innovation in the Nordic Countries' (MEPIN) project (Bloch 2010a,

2010b, 2013; Bloch and Bugge 2013) which developed a definition for use in measuring innovation in the public sector. This was an important project as it led to a proposal in 2014 to the OECD Working Party of National Experts on Science and Technology Indicators (NESTI) for the development of a manual, like the *Oslo Manual*, that would provide guidance on measuring and interpreting data on public sector innovation. The timing was not ideal as the revision of the *Oslo Manual*, leading to the fourth edition, was scheduled to begin in 2015, the year in which the seventh edition of the *Frascati Manual* (OECD 2015c) was published.

At the same time, there was a project in Finland to measure innovation by household sector and individuals. The project was led by Jari Kuusisto for Finland and Eric von Hippel led the consultants' team (Kuusisto et al. 2013; de Jong et al. 2015). Part of the motivation was to find a culture of innovation that could support household innovation and the growth of the innovators into large businesses providing jobs and economic growth. This was a successful project but the activities being measured did not align with the definition of innovation in the third edition of the *Oslo Manual*. Individuals and households could develop new or significantly improved products but introducing them on the market was not yet a consideration.

The discussion of this problem led to a paper, Gault (2012), in which a small change to the *Oslo Manual* definition of innovation was proposed. The proposal was to replace 'introduced on the market' for a product innovation with 'made available to potential users' in paragraph 150 of OECD/Eurostat (2005). If this was done, the new or significantly improved product developed by the household, or the individual, was a product innovation if it was made available to potential users. From the perspective of official statistics, there were three options for the innovator to make the product available. The product could be made available to

1. a producer, demonstrating the improvement to the product;
2. an existing business or a new business of the innovator;
3. a community of practice or a peer group.

The first two options would be captured in a CIS if the producer, or the innovator's business, had ten or more employees. The third option would not appear in official statistics, but it was a means of including households, or individuals, as innovators. A more detailed discussion of user innovation and official statistics is found in Gault (2016).

It was also evident that such a change to the definition could apply to the measurement of innovation in the public sector and this led to the broader question of whether there could be a definition of innovation applicable in all economic sectors. A discussion paper was provided to the team working (Gault 2015) on the revision of the *Oslo Manual*, leading to the fourth edition.

This historical review demonstrates that the definition of innovation, for measurement purposes, is still evolving, as are the measurement activities that are governed by the definition.

4.7   CONCLUSION

This chapter has reviewed the means for monitoring and evaluating implemented innovation. These include country reports, based on innovation surveys, country reviews conducted by experts on innovation policy, the use of scoreboards for ranking of countries. All of these can be used to monitor and evaluate the state of innovation in the business sector in participating countries. Public sector innovation was introduced to set the stage for discussion, in Chapter 7, of a general definition of innovation applicable in all economic sectors.

NOTES

1. See https://www.gov.uk/government/statistics/uk-innovation-survey-2017 -main-report (accessed 17 March 2020).
2. See https://www150.statcan.gc.ca/n1/daily-quotidien/190326/dq190326b -eng.htm (accessed 17 March 2020) and references provided there.
3. See https://www.oecd.org/innovation/oecd-reviews-of-innovation-policy .htm (accessed 17 March 2020) for recent examples.
4. An example is a review of Viet Nam done with the OECD. https:// www.worldbank.org/en/country/vietnam/publication/a-review-of-science -technology-and-innovation-in-vietnam (accessed 17 March 2020).
5. An example is Tanzania, see https://www.unido.org/sites/default/files/2011 -04/Tanzania_0.PDF (accessed 17 March 2020).
6. This is described in https://en.unesco.org/go-spin (accessed 17 March 2020).
7. See http://www.oecd.org/sti/inno-stats.htm (accessed 17 March 2020).
8. The use of composite indicators is discussed in OECD/Eurostat (2018), para. 11.23 and Box 11.2.
9. 'Process' here is defined in the third edition of the *Oslo Manual* (OECD/ Eurostat 2005), not in the fourth edition (OECD/Eurostat 2018) where 'process' is different.

10. See https://ec.europa.eu/growth/industry/innovation/facts-figures/regional _en (accessed 17 March 2020).
11. Business sector innovation was discussed for about 15 years before there was agreement on what could be accepted as a definition of innovation for measurement purposes, leading to the first edition of the *Oslo Manual* (OECD 1992). See Chapter 6 of this book.
12. See https://op.europa.eu/en/publication-detail/-/publication/fe2a3b4b -3d7e-444d-82bc-790a0ab33737 and https://ec.europa.eu/growth/content/ innovation-public-sector-0_en (both accessed 17 March 2020).

5. Developing innovation policy

5.1 INTRODUCTION

In Part II, so far, selected innovation policies were reviewed in Chapter 3 and ways of monitoring or evaluating innovated policies were discussed in Chapter 4. In both cases, and in this chapter, innovation policy is focused on the business sector. The question now is how innovation policy has been developed. This can occur as a result of the evaluation of existing policies, such as those in Chapter 3, with a view to improving the outcomes. An alternative approach is the development of an entirely new innovation policy to address new objectives.

In the business sector, innovation policy, initiated by government, is expected to deliver good jobs, economic growth and other important outcomes such as wellbeing. How policy is designed to deliver these objectives is an issue and the approach is not the same in all countries. In a previous discussion of developing innovation policy (Gault 2010), the distinction was made between countries that took 'a whole of government' approach and those that had fragmented policies which varied from ministry to ministry. As an example, one ministry could promote R&D performance while another promoted the use of information and communication technologies (ICTs). There are overlaps which may cause problems if the ministries do not communicate with one another.

Innovation policy can also be part of a regional development policy, promoting innovation in developing regions (European Commission 2019) in the hope that the underdeveloped regions can catch up with the more prosperous ones. There is sectoral innovation policy, promoting innovation in agriculture, ICTs, bio- and nano-technology, artificial intelligence, or other technologies that are giving rise to the 'fourth industrial revolution' (Schwab 2017, 2018). There is 'restricted' innovation which is used to identify a measurement of innovation with a particular characteristic, such as inclusiveness or sustainability.

A decade ago, countries were trying to deal with the consequences of the financial crisis of 2008 and policy makers were looking to innovation policy for ways out of the crisis and to support recovery. At the time, there was limited discussion of the cause of the crisis being the introduction on the market of new or significantly improved financial products, based on sub-prime mortgages, as there was a belief, still prevalent, that all innovation is good. In that period, there was little discussion of innovation as a systems phenomenon outside of the scholarly community (Edquist 2005). A systems approach to innovation policy (Chapter 2) includes direct intervention to change the behaviour of firms, indirect intervention to change the framework conditions that influence innovation in the firms (trade regulation, rules governing employment, higher education outcomes …) and, of considerable importance, policy intervention to change the flows of knowledge, material and human resources into and out of firms. This is not a criticism of research on national innovation systems,[1] but a more general observation about the use, or lack of use, of a basic systems approach to describe innovation in firms, the innovation activities engaged in by the firms, linkages with other economic institutions and the outcomes and impacts of innovation policy, once implemented.

Edler and Fagerberg (2017) separate innovation policy into three categories. The first, citing Ergas (1986), is mission-oriented innovation policy such as DARPA's creation of the internet which has had significant economic impact (Mazzucato 2013). The purpose of the implemented policy was to solve a problem and the work was done by a public sector organisation. Public sector innovation is an important issue which will be considered in greater detail in Part III and global challenges, more specifically, in Chapter 9.

The second is invention-oriented innovation policy driven by R&D, which has been referred to as R&D, research, or science policy as well as innovation policy. This is an issue in countries with a small private sector where governments are more inclined to create and support research institutions and university departments.

The third category is system-oriented policy which was introduced in Chapter 2, along with National Innovation Systems. This book takes the view that all innovation policies need to take a systems approach as innovation is a systems phenomenon (Chapter 2).

Since the financial crisis of 2008 there have been major changes in the activity of firms as the economy has become digital. That was not an issue in 2008, but it is now, and there are implications for policy (Guellec and Paunov 2018) and statistical measurement (OECD 2019a) to support policy development.

5.2   GOVERNING INNOVATION POLICY

In Chapters 2 and 3, the point was made that innovation is a systems phenomenon and innovation policy development must take that into account if it is to be effective when it is implemented. Borrás and Edquist (2019: 232) argue the case for 'holistic innovation policy' which includes all parts of the innovation system. The OECD Innovation Strategy (2015a) notes the utility of 'a whole of government' approach to innovation policy. This was also discussed in Gault (2010) from an operational perspective dealing with the policy instruments that can be used to implement an innovation policy, whether a partial one in a single ministry or a whole of government approach.

For there to be a whole of government approach to innovation policy, the lead must come from a high level. In some countries this can work, but in others not. Consider the review of innovation policy and indicators in the US provided by Christopher Hill (2013). John Marburger, at the OECD Blue Sky Forum-2, addressed the whole of government approach as follows:

> Overall science planning and policy making is accomplished through a bewildering variety of advisory panels, interagency working groups, and Executive Branch policy processes, the most important of which is the annual budget process that synthesizes the proposal presented annually by the President to Congress. In Congress, multiple committees and subcommittees authorize and appropriate funds in an intense advocacy environment from which politics is rarely excluded. (Marburger 2007: 30)

Whether the policy is whole of government or limited to one ministry, once it has been implemented statistical measurement may be considered so that the policy can be monitored and, eventually, evaluated. This would support policy learning leading to changes to the policy to make it more effective.[2]

For a number of reasons, the governance of innovation policy has changed over the last decade and Edler and Fagerberg (2017) have provided a review of innovation policy which addresses: what innovation policy is; why there is an innovation policy, with a discussion of theoretical rationales; and innovation policy in practice.

5.3 MEASURMENT AND RESTRICTIONS

This book is about statistical measurement of innovation and in Part II, the focus is on innovation in firms and the Community Innovation Survey (CIS) and CIS-like surveys are the measurement examples. This section deals with methodological constraints and restrictions which are, or could be, part of the measurement. Restrictions are important as they help to respond to policy questions that go beyond an interest in the propensity of firms to innovate.

An example of a methodological constraint on a survey is the three-year reference period for observing innovation in CIS. Consider the following question, 3.1, from the CIS 2018 survey:

> During the three years 2016 to 2018, did your enterprise introduce any:
> New or improved goods   Yes/No
> New or improved services Yes/No

The question, 3.1, is prefaced with: 'A product innovation is a new or improved good or service that differs significantly from the firm's previous goods or services and which has been implemented on the market'. This differs from the definition of innovation in the 3rd edition of the *Oslo Manual* (OECD/Eurostat 2005), and from the definition of business sector innovation in the 4th edition (OECD/Eurostat 2018). This reflects the fact that the question was drafted in the middle of three years of revision of the *Oslo Manual*. That is not an issue, but the constraint, 'During the three years 2016 to 2018' is. This is also discussed in the *Oslo Manual* (OECD/Eurostat 2018: Chapters 3, 9 and 11).

A firm that introduced on the market a new or improved product on 31 December 2015 is not an innovative firm in the CIS 2018 unless there is a new innovation within the period 2016–18. This can be a concern to firms that have invested in the new product and brought it to market only to find that they cannot report this in their response to the questionnaire as the innovation was out of scope. If the firm was in the sample for the CIS 2016, the firm would be recorded as an innovative firm, but not for CIS 2018.

In addition to methodological constraints, there are restrictions that may be imposed on the measurement of innovation as they are of policy relevance. The minister is not looking for the propensity of firms to innovate but the influence of innovation policy on key issues of the day. Reporting on the likelihood of innovation to advance sustainability and inclusion are examples, so are jobs and growth. They will be introduced here and discussed further in Chapter 9.

A first requirement is that there are definitions of sustainability and inclusion and of the population to be studied. Is the issue of sustainability economic, environmental or social or a combination of all three? Once those questions are answered, survey questions can be drafted, questionnaires designed, and measurement made. The same applies for inclusion. What is to be included (the poor, poorly served social groups, adherents of other cultures, women, …)?

There are two approaches to measuring these innovation restrictions (Gault 2018a). The first is the provision of a product innovation and the impact on the users of that product. To determine a change in inclusion or sustainability the users must have been surveyed and these could be surveys of any economic sector. For household products, social surveys would be required. The second approach is process innovation and the resulting impact – is it inclusive and sustainable and is there a causal link between the process innovation and the perceived outcome?

5.4   INTERNATIONAL INNOVATION POLICY

Innovation policy can come from international or supranational organisations. The OECD Innovation Strategy (OECD 2015a) has already been discussed and the importance of 'a whole of government' approach. Since 2015 there has been a rapid change to the digital economy and the OECD's Digital and Open Innovation Project has produced reports on the digital economy (OECD 2017e), innovation policy (Guellec and Paunov 2018; OECD 2019a) and measurement (OECD 2019d). A point to note is that the OECD provides working examples of innovation policy, but it does not propose innovation policies for Member Countries.

The European Commission provides an approach to innovation policy in a number of domains not all of which are covered by the third edition of the *Oslo Manual* (OECD/Eurostat 2005). They are:[3]

- social innovation
- demand-side innovation policies
- public sector innovation
- workplace innovation.

It is evident that the European Commission sees these topics as significant, even if the measurement mechanisms, including definitions, are not present for all of them to support monitoring and evaluation.

There is also the Enhanced European Innovation Council (EIC) pilot.[4] It brings together the parts of Horizon 2020 that provide funding, advice and networking opportunities for entrepreneurs and innovators.

In a third example, the African Union (AU 2014) has published a Science, Technology and Innovation Strategy for Africa (STISA – 2024). It expects Member States to implement the strategy, to agree on indicators and to report on the findings of statistical measurement. STISA – 2024 is part of a longer term initiative, the Agenda 2063.

5.5 CONCLUSION

In this chapter approaches to developing innovation policies were reviewed from the country level and from that of international/supranational organisations, the African Union, the European Union and the OECD.

The focus has remained on innovation in the business sector, but with some reference to the need to measure innovation, viewed as a systems phenomenon, in all economic sectors.

The principal change in the economy in the last decade is digitalisation which has been introduced. What has not been discussed, now that there is a clear need for measuring innovation in all sectors, is how to define innovation to support policy making and measurement. That follows in Part III.

NOTES

1. Chaminade et al. (2018) provides a thorough review of the subject. See also Soete et al. (2010).
2. The reader may find that not all ministries support monitoring or evaluation of their policy interventions.
3. See https://ec.europa.eu/growth/industry/innovation/policy_en (accessed 17 March 2020).
4. See https://ec.europa.eu/research/eic/index.cfm (accessed 17 March 2020).

PART III

Measuring innovation

6. Defining innovation for measurement purposes

6.1 INTRODUCTION

The definition of innovation for measurement purposes has a long history.[1] For OECD and EU countries a definition was agreed in 1992 when, after years of experimentation with innovation surveys, the knowledge gained was codified in the first *Oslo Manual* (OECD 1992). It dealt with product and process innovation in manufacturing. In 1997 (OECD/Eurostat 1997) the manual was revised to include marketed services, construction and utilities, in addition to manufacturing (OECD/Eurostat 1997: para. 15). Product innovation remained the same, but process innovation had product delivery added to its definition.

Between the release of the second edition in 1997 and the third in 2005, there was an innovation project in Latin America and the Caribbean which led to the *Bogotá Manual* (RICYT/OEC/CYTED 2001). The manual focused on manufacturing and differed from the *Oslo Manual* in its treatment of local capabilities and capacity building and its discussion of organisational change and market development. This influenced the third edition of the *Oslo Manual* (OECD/Eurostat 2005) which added both organisational change and market development to the definition of innovation and an Annex, at the request of RICYT, on interpreting the manual for use in innovation surveys in developing countries (OECD/Eurostat 2005: 135). This was a step towards making the *Oslo Manual* accessible globally. Another step, in the third edition, was the chapter on linkages. The innovation activities of a firm depend on links to sources of information, knowledge, technologies, practices and human and financial resources. In a systems approach (Chapter 2) to the measurement of innovation, linkages matter more than firms. Linkages, such as value chains, and their impacts, tend to be different in developing and developed countries.

A common characteristic of the first three editions of the *Oslo Manual* is that they all provide guidance for measuring firm-level innovation in the business sector. This is not surprising as innovation in the business sector is seen as a primary source of wealth creation and economic growth. With each edition, the sector coverage broadened, and the third edition included innovation in the entire business sector (OECD/Eurostat 2005: para. 27). This was the first appearance in the *Oslo Manual* of primary industries (agriculture, forestry, fishing and mining).

Starting from manufacturing and moving through three editions to the entire business sector was a natural progression resulting from learning by doing and responding to policy needs that did not deal only with manufacturing. A second progression, still evolving, followed the first OECD Blue Sky Forum in 1996. It was the growing emphasis on a systems approach to measuring innovation. A third progression was the recognition that innovation did not happen only in the business sector. The first and second editions of the *Oslo Manual* noted that 'Innovation can of course occur in any sector of the economy, including government services such as health or education' (OECD 1992: para. 84; OECD/Eurostat 1997: para. 15). The same statement appeared in the third edition with a minor change ('of course' was dropped) (OECD/Eurostat 2005: para. 27). However, motivated by findings from measuring innovation in other sectors (Gault 2018a), the fourth edition provided a general definition of innovation applicable in all economic sectors. This was a major step in defining innovation for measurement purposes. The fourth edition of the *Oslo Manual* (OECD/Eurostat 2018: para. 102) makes the point that:

> specific innovations can involve the participation of multiple actors across sectoral boundaries. These units can be linked through various methods, such as funding mechanisms, hiring of human resources, or informal contracts.

This is discussed further in Section 6.2.

In parallel with the work on innovation in all sectors, policy developers were showing interest in 'restricted' forms of innovation (Gault 2018a). In the business sector, surveys could be conducted according to *Oslo Manual* definitions, using a statistical sample, and identifying a sample population of innovative firms. This would provide the basis for a population estimate of the propensity to innovate. What it did not provide was information on whether the innovation was sustainable, inclusive, pro-poor, green or any combination of those or any other constraints that were important to policy makers. Dealing with this is discussed in Section 6.3.

Deciding not to have an Annex interpreting the manual for use in developing countries in the fourth edition was noted as a step towards a global manual.[2] In the fourth edition the whole manual was accessible to developing and developed countries. This brought the *Oslo Manual* into line with the *System of National Accounts, 2008* (*2008 SNA*) (EC et al. 2009: para. 1.4) where it is noted that:

> There is no justification, for example, for seeking to define parts of the SNA differently in less developed than in more developed economies, or in large relatively closed economies than in small open economies, or in high-inflation economies than in low inflation economies.

Just as innovation does not happen in isolation, neither does the development of an innovation manual. In parallel with the work leading to the fourth edition of the *Oslo Manual*, the International Organization for Standardization (ISO) was developing a definition of innovation for the purpose of innovation management in its ISO 56000 series. To ensure compatibility, OECD and ISO worked closely together during the period of revision (Gault and Hakvåg 2018). This is discussed in Section 6.6.

6.2 DEFINITION OF INNOVATION IN THE *OSLO MANUAL*

The fourth edition of the *Oslo Manual* has two definitions of innovation: a general definition applicable in all economic sectors and a sector-specific definition for the business sector.[3] Both are presented and then there is a discussion of the measurement implications of the two definitions.

The general definition of innovation in the fourth edition of the *Oslo Manual* is the following.

> An **innovation** is a new or improved product or process (or combination thereof) that differs significantly from the unit's previous products or processes and that has been made available to potential users (product) or brought into use by the unit (process). (OECD/Eurostat 2018: Chapter 1, para. 1.25, Chapter 2, para. 2.99)

The definitions of sectors are those of the SNA (EC et al. 2009). 'Unit', in the general definition, refers to the 'institutional unit' in the sector and examples are provided in Chapter 7 of this book. In the business sector, the 'institutional unit', or 'unit', is the firm.

In principle, the general definition could be applied to all sectors including the business sector. However, the *Oslo Manual* has, from its beginning in 1992, been a manual for collecting, reporting and using data on innovation in the business sector where a product innovation is introduced on the market and a process innovation is brought into use. This has continued in the fourth edition. The definition of innovation in the business sector is close to that in the third edition, allowing continuity in innovation surveys, including the EU Community Innovation Survey (CIS), which has been conducted since 1992. From 2006, CIS has occurred in EU Member States, and other states, every two years.[4]

The definition of business innovation is:

> A **business innovation** is a new or improved product or business process (or combination thereof) that differs significantly from the firm's previous products or business processes and that has been introduced on the market or brought into use by the firm. (OECD/Eurostat 2018: Chapter 3, para. 3.9)

The definitions of innovation in the third and fourth editions of the *Oslo Manual* are similar. This has the advantage that statistics gathered through surveys based on the *Oslo Manual* require little additional guidance of respondents. The similarity is illustrated in what follows.

The definition of innovation in the third editions is:

> An **innovation** is the implementation of a new or significantly improved product (good or service), or process, a new marketing method, or a new organizational method in business practices, workplace organization or external relations. (OECD/Eurostat 2005: para. 146)

For the definition to be complete, 'implementation' must be defined. The definition of 'implementation' is found in the following:

> A common feature of an innovation is that it must have been **implemented**. A new or improved product is implemented when it is introduced on the market. New processes, marketing methods or organisational methods are implemented when they are brought into actual use in the firm's operations. (OECD/Eurostat 2005: para. 150)

Comparing the third and fourth edition definitions, they are the same for product innovation. In the case of a process innovation, the third edition has three functional categories, the fourth edition has six (OECD/

Eurostat 2018: Table 3.1). The correspondence of the three categories in the third edition to the six categories in the fourth edition is more indicative than precise, but the advantage of the six is that they support better measurement of the innovation process. The categories follow.

The first category of the third edition is 'process innovation', the implementation of a new or significantly improved production or delivery method (OECD/Eurostat 2005: para. 163). In the fourth edition, two categories are used: 1. Production of goods and services and 2. Distribution and logistics.

The second category in the third edition is marketing innovation, the implementation of a new marketing method involving significant changes in product design or packaging, product placement, product promotion or pricing (OECD/Eurostat 2005: para 169). This corresponds in the fourth edition to the category: 3. Marketing and sales.

The third category in the third edition is organisational innovation (OECD/Eurostat 2005: para. 177), where an organisational innovation is the implementation of a new organisational method in the firm's business practices, workplace organisation or external relations. The closest correspondence in the fourth edition is: 5. Administration and management. There are two other categories in the fourth edition: 4. Information and communication systems and 6. Product and business process development.

Turning to the fourth edition definition, the only difference of substance between the definition of business innovation and the general definition is 'made available to potential users' for a product compared with 'introduced on the market'. The consequences of using the general definition in the business sector are explored in Section 6.4 and in Chapter 7. It appears here in a chapter on definitions because of the presence of the digital economy and the implications the digital economy has for definitions of innovation.

6.3   RESTRICTED INNOVATION AND MEASUREMENT

The *Oslo Manual* provides guidelines for collecting, reporting and using data on innovation. It does not deal with restrictions on the type of innovation to be measured or on its outcomes. Examples already mentioned are sustainable innovation,[5] inclusive innovation,[6] green innovation and pro-poor innovation. There are others.

With a growing policy interest, there is a need for measurement of restricted innovation in order to support monitoring and evaluation of policies designed to promote the restricted innovation. However, in most cases, the desired objective of sustainability or inclusion cannot be determined by one measurement. The first measurement serves to provide a baseline, but others must follow to confirm that the result of the innovation is sustainable, or that the excluded population is included. Where people, and their lives, are concerned the subsequent measurements may include social as well as business surveys.

Also, for the measurement to be successful there must be a definition of 'sustainable' or 'inclusion', as well as for any other term used to restrict the innovation. The time interval, or intervals, to confirm that the object is achieved must also be agreed.

An example of restricted innovation dealing with sustainability and inclusion is provided by Mashelkar (2012, 2014):

> Inclusive innovation is any innovation that leads to affordable access of quality goods and services creating livelihood opportunities for the excluded population, primarily at the base of the pyramid and on a long term sustainable basis with a significant outreach.

In this definition, 'any innovation' links to the *Oslo Manual* definition of innovation and the restrictions follow, along with at least two time scales, the initial measurement and a subsequent one to confirm sustainability and other restrictions in the definition.

This example is discussed at greater length in Gault (2014, 2018a), which leads to a fundamental question: Are governments willing to pay for the new indicators that are needed to monitor and evaluate the relevant innovation policies?

Sustainability requires measurement over at least one time period to demonstrate that the objective, however defined, has been achieved. Inclusion, again depending on how it is defined, can be built into the innovation process. A government procurement call, for example, could require that firms that apply have a labour force with at least a specified percentage of the excluded population. However, in such a situation, firms that met the requirement some time ago could apply and others not. The policy objective of increasing the employment opportunities for the excluded population would not be achieved. This is discussed further in Chapter 9.

A final point on the definitions of innovation in this chapter is that they include no moral or ethical conditions. Nowhere is there an assumption that innovation is good, or bad. As discussed above, the definition of innovation can be restricted to deal with inclusive, social or sustainable innovation, the restricted definition must still support statistical measurement through surveys or other means of data collection, leading to indicators which can be used to support policy development, and the monitoring and evaluation of implemented policy. To make this happen, the terms used in the restricted definition must be defined.

However, definitions do not always support statistical measurement. Definitions can also be used to guide discussion and theorising, and this is an important means of developing a subject. The measurement approach and the theorising approach[7] are not at variance but are complementary.

6.4 DIGITAL ECONOMY, INNOVATION AND MEASUREMENT

The digital economy is everywhere and so is innovation. The general definition of innovation in Section 6.2 is applicable in all the economic sectors (discussed in Chapter 7), including the business sector. The discussion starts with product innovation in the business sector but uses the general definition.

The general definition imposes two conditions on a product for it to be a product innovation. The product

• is new or improved and differs significantly from the unit's previous products and
• has been made available to potential users.

Nowhere is the product required to be sold at economically significant prices, it needs only to be 'made available to potential users', and to have satisfied the first condition. The product can be given away. This raises a question about why any firm would give away a product.[8]

There are many cases where products are provided to potential users at zero price. These include email accounts, cloud services and apps. They may be given freely to potential users, but there is an expectation that the producer is connected to the user allowing the producer to follow what the user is doing. This may seem like a barter transaction, but how many users think of it this way? The rationale is that the service provided is improved if information on transactions, processing and searching can be collected and used by the producer. This suggests that, in a digital economy, with digital products at zero price, there is a de facto third condition. The product connects the user to the producer.

This is not found in the *Oslo Manual*, but it provides a basis for examining zero price product innovations and their impact on users. Users are not just households and individuals, but actors in all economic sectors of the SNA. This will be discussed in Chapter 7.

The policy implications of digital innovation are discussed in Chapter 10 (OECD 2019a, 2019d).

6.5   *OSLO MANUAL* DEFINITIONS AS INTERNATIONAL STANDARDS

In the introduction to this chapter, there was reference to the *Oslo Manual* becoming an international standard. The fourth edition is designed to be accessible to developing countries as well as to the 36 OECD Member Countries, observer countries and 28 EU Member States (in January 2020). The move towards being a global standard started in the revision process with broadening of the industrial coverage until all industries in the business sector, including agriculture, forestry, fishing and mining, were covered by the manual.

The informal economy and the informal sector are present everywhere, but they are more relevant in developing countries. This is discussed further in Chapter 8. Measuring innovation in the informal economy is a challenge, but a necessary one if the functioning of the economy is to be understood, so that innovation policy can be developed (Kraemer-Mbula and Wunsch-Vincent 2016).

An advantage of having an international standard is that it is maintained by an organisation, in this case the OECD with Eurostat, which manages the cost of revisions and the research needed to keep the standard current. This was part of the reason why the African Union/NEPAD, at the meeting of the first Intergovernmental Meeting on Science, Technology and Innovation Indicators (Gault 2010: 139) in Mozambique, in 2007, adopted the *Frascati Manual* (OECD 2002) and the *Oslo Manual* (OECD/Eurostat 2005) as standards for R&D and innovation surveys in Africa. These surveys led to three editions of *African Innovation Outlook* (AU-NEPAD 2010; NPCA 2014; AUDA-NEPAD 2019) and discussion of policy related to science and technology, and to innovation, in Africa.

To support a more inclusive and ongoing dialogue, the African Union/ NEPAD was invited in 2007 to send an observer to OECD Working Party of National Experts on Science and Technology Indicators (NESTI) to join the observers from RICYT in Latin America and the Caribbean, the UNESCO Institute for Statistics, and other observers from non-member countries. This resulted in contributions to discussions during the revision of the *Frascati Manual* (OECD 2015c) and the *Oslo Manual* (OECD/Eurostat 2018) which helped make the manuals international standards.

6.6 INNOVATION AND INTERNATIONAL STANDARDS

Standards influence innovation. They appear in rules governing activities in countries, in international treaties, including those dealing with trade and intellectual property protection, and practices (Hawkins et al. 2017). They can be legally binding or de facto standards that communities of practice accept.

The ISO work on a standard for innovation management in its 56000 series is discussed here because it illustrates the ability of two international organisations to collaborate for the common good.

The example is the understanding of 'value'. The approach in the *Oslo Manual* is that adding value can be an intention of innovation but when it comes to measurement, the outcome may not have added value. The new car model may not sell, the pharmaceutical product may have unexpected and undesired outcomes and the digital product may undermine the privacy of the individual user. So long as the products are 'new or improved' and they have been 'introduced on the market' (or made available to potential users), they are innovations.

In the OECD-ISO discussions a qualifying sentence was added to the section on value. It was 'Value can be e.g. created, realized, acquired, redistributed, shared, lost, or destroyed.' This is a single illustrative example from productive discussions between the two international organisations. More can be found in Gault and Hakvåg (2018).

The work with the ISO is an example of cooperation of international organisations to ensure that their definitions are not incompatible. This raises a question about whether there should be more collaboration among international and supranational organisations to ensure that terms critical to statistical measurement and policy development, monitoring and evaluation, be used and understood in the same way.

6.7   CONCLUSION

This chapter has reviewed the general definition of innovation provided in the fourth edition of the *Oslo Manual* (OECD/Eurostat 2018) and earlier work on general definitions (Gault 2018a) applicable in any economic sector of the SNA (Chapter 7).

While the general definition is presented in the fourth edition of the *Oslo Manual*, the manual has been, and still is, a guide to measuring innovation in the business sector. For that reason, the definition of innovation in the business sector has been introduced as a special case of the general definition. This allows a connection with what has gone on in earlier *Oslo Manuals* while maintaining the requirement that a product must satisfy a number of conditions in order to be a product innovation, including being introduced on the market.

While the guidance on measuring product innovations in the business sector in the fourth and earlier editions of the *Oslo Manual* is well established and understood, there is no comparable experience of changing the condition that a product be introduced on the market to being made available to potential users. Once this change is made, the implicit condition that a product introduced on the market is available at a market price is gone. Product innovations may be made available to potential users at other than economically significant prices, including zero prices. This is particularly relevant to measurement of innovation in the digital economy as there are many product innovations provided to potential users at a price of zero.

The digital economy is growing rapidly, supported by readily accessible technology. It is changing the way in which people interact with the economy and how service providers and users innovate. Given the rate of change, understanding innovation in the digital economy is a priority for the 2020s and for incorporating the outcomes into the SNA.

Digitalisation and innovation are discussed in the fourth edition of the *Oslo Manual* (OECD/Eurostat 2018: Section 1.2.5) where examples are provided to illustrate that digitalisation can be 'an innovation process in its own right and a key factor driving innovation'. Digitalisation can also support measurement of innovation which may not be intended as statistical measurement. Examples of this are provided in (OECD/Eurostat 2018: para. 1.53).

Restricted innovation was examined, noting that, in some cases, a single measure of innovation may not capture inclusive, or sustainable, or green innovation. Once the definition of innovation is restricted, and a baseline survey conducted, there may be a requirement for additional surveys to demonstrate the progress of inclusion, sustainability or growing green innovation. Subsequent surveys may be directed at people as well as at institutions.

Standards are relevant to innovation measurement and to international comparability of findings. The *Oslo Manual* has progressed since the first edition in 1992 from being focused on developed countries and limited in its coverage to full coverage of all institutional units in all economic sectors of the SNA in 2018. The importance of being a de facto international standard for measuring innovation and related innovation activities lies in the country comparisons that result from measurement that adheres to the standards. This was discussed in Chapter 4.

While the *Oslo Manual* is now an international standard, there are other international standards that influence the activity of innovation, innovation activities and framework conditions that govern the behaviour of institutional units. These influence the discourse on the subject of innovation which has a number of approaches to understanding innovation and its impact.

Now that the activity of innovation has been defined, and relevant factors considered, Chapter 7 introduces the SNA and the economic sectors where innovation happens.

NOTES

1. For more detailed discussions of the history of the definition of innovation for measurement purposes, see Smith (2005) and Gault (2013).
2. Appropriately, the proposal for the removal of the Annex came from an observer from the African Union/NEPAD at the 2016 meeting of NESTI. It was approved by delegates.
3. It is important to note that the definition of innovation in the business sector is a subset of the general definition of innovation.
4. See https://ec.europa.eu/eurostat/web/microdata/community-innovation -survey (accessed 17 March 2020).
5. The Sustainable Development Goals (SDGs) and the role of innovation in achieving them are discussed in Chapter 9.
6. See Nesta publications for discussions of inclusive innovation (https://www .nesta.org.uk, (accessed 17 March 2020).
7. See George et al. (2019) and the definition of inclusive innovation.
8. The author is indebted to Nordine Es-Sadki from UNU-MERIT for sharing the reaction of respondents from firms when asked about zero price products. That gave rise to the third condition for a product innovation that is not made available at an economically significant price.

7.   Measuring innovation in all economic sectors

7.1   INTRODUCTION

Chapter 6 presented a general definition of innovation applicable in all economic sectors and an application of the general definition to the business sector. Chapter 7 now provides definitions of the economic sectors, as specified in the *System of National Accounts, 2008* (*2008 SNA*) (EC et al. 2009). Also provided are examples of statistical measurement of innovation in the sectors, where they are available, and suggestions where they are not. In the case of the business sector, consideration is given both to the sector-specific definition used in the *Oslo Manual* (OECD/Eurostat 2018: para. 3.9) and to the general definition.

Once the economic sectors are defined, the focus of the chapter is on measurement. As discussed in Chapter 6, affirmative responses to two survey questions are required to confirm the presence of innovation. The first is a question asking if there is a 'new or improved product or process (or combination thereof) that differs significantly from the unit's products or processes'. The response to this question requires a judgement on the part of the respondent. The second question does not. It asks what is done by the unit with that product or process. Has the product been 'made available to potential users'? Or, has the process been 'brought into use by the unit'? What is done with the product or process is observable and it either has happened or not. It is not a matter of judgement. A survey statistician can, from the answers to these two questions, infer the presence of innovation.[1]

7.2   THE SYSTEM OF NATIONAL ACCOUNTS

The System of National Accounts (SNA) serves as a coordinating framework for data collection and indicators that support the understanding of the economy and the development and monitoring of economic policy

(EC et al. 2009: para. 1.57). The best known indicator in the SNA is the Gross Domestic Product (GDP).[2]

The SNA provides a flow of information that can be used for monitoring, analysis and evaluation of data over time. It also deals with links of the economy to the rest of the world (EC et al. 2009: para. 1.2). Directly relevant to measuring innovation in all economic sectors is the ability to 'observe and analyse the economic interactions taking place between different sectors of the economy' (EC et al. 2009: para. 1.3). The SNA supplies a conceptual framework for ensuring the consistency of the definitions and classifications used in different, but related, fields of statistics. It also acts as an accounting framework to ensure the numerical consistency of data drawn from different sources in the system (EC et al. 2009: para. 1.57).

From the first edition of the *Oslo Manual* (OECD 1992), the conventions of the SNA have been used, the most common of which is 'product' to refer to goods and services (EC et al. 2009: para. 2.36). In the fourth edition of the *Oslo Manual* (OECD/Eurostat 2018), Chapter 2 introduces the SNA, and its use in the manual. The term 'unit' in the general definition of innovation refers to the 'institutional unit' in economic sectors. The definition of an 'institutional unit' is found in the *2008 SNA*, para. 4.2:

> An institutional unit is an economic entity that is capable, in its own right, of owning assets, incurring liabilities and engaging in economic activities and in transactions with other entities.

The SNA (EC et al. 2009: para. 4.24) assigns all residential institutional units to one of five sectors, the:

* non-financial corporations sector
* financial corporations sector
* general government sector
* non-profit institutions serving households sector (NPISHs)
* households sector.

For use here, and to conform with the *Oslo Manual*, the first two sectors are combined to become the 'business sector' (see more in the next section). For the fifth sector, 'Households' is replaced with 'household'.[3]

Innovation in the public sector is an area of interest and the SNA provides the following definition. 'The public sector includes general government and public corporations.' However, some effort is required to decide where non-profit institutions are placed, and the same question arises for institutions which may be public institutions or are part of general government. This is discussed in the *2008 SNA* (EC et al. 2009: 436, Defining the general government and public sectors).

The *2008 SNA* is the most recent in a series of SNA manuals produced by international and supranational organisations and used by countries to guide their SNA and the production of key indicators such as GDP. Decisions taken by the SNA community have an impact on the measurement of innovation and of innovation activities. An example is the decision to capitalise expenditure on software development, making it a capital investment and no longer an expense (EC et al. 1994: 648). In the following revision of the SNA, the Canberra II Group was responsible for the recommendation to capitalise R&D. In the course of the deliberations, members of Canberra II and NESTI met twice, once in Berlin in 2006 and once in Paris in 2007. In the 2008 revision, R&D was capitalised[4] (EC et al. 2009: xlviii).

From the perspective of measurement, the SNA has capitalised two out of the eight business innovation activities (EC et al. 2009: para. 1.36). The remaining six are: engineering, design and other creative work activities; marketing and brand equity activities; intellectual property (IP)-related activities; employee training activities; activities related to the acquisition or lease of tangible assets; and innovation management activities. Haskel and Westlake (2018) provide a relevant discussion of 'the rise of the intangible economy'.

This introductory section is deliberately short, but it connects the terms in the definitions of innovation (product and unit) to the *2008 SNA*, links the four economic sectors in the *Oslo Manual* to the five sectors in the SNA and it defines the public sector. More high-level information is found in Chapter 2 of the *SNA Manual 2008* and a more detailed discussion of institutional units and sectors is found in Chapter 4 of the same manual. Finally, the link is made between business innovation activities in the fourth edition of the *Oslo Manual* and the *2008 SNA*.

7.3   BUSINESS SECTOR

The business sector, here and in the *Oslo Manual*, is the combination of the SNA non-financial and the financial corporations sectors. They are defined as follows:

> **Non-financial corporations** are corporations whose principal activity is the production of market goods or non-financial services. (EC et al. 2009: para. 4.94)

> **Financial corporations** consist of all resident corporations that are principally engaged in providing financial services, including insurance and pension funding services, to other institutional units. (EC et al. 2009: para. 4.98)

When these are combined, they cover all resident corporations that provide goods or services.

7.3.1   Business Innovation and the Market

In the fourth edition of the *Oslo Manual*, processes, marketing methods and organisational methods of the third edition are combined into processes with six sub-categories (OECD/Eurostat 2018: Table 3.1), discussed in Chapter 6, Section 6.2. The definition of business innovation is the following:

> A **business innovation** is a new or improved product or business process (or combination thereof) that differs significantly from the firm's previous products or business processes and that has been introduced on the market or brought into use by the firm. (OECD/Eurostat 2018: para. 3.9)

As with the previous editions of the *Oslo Manual*, for a product to be a product innovation, it must be 'introduced on the market' with the implication that the product is offered at economically significant prices (EC et al. 2009: para. 22.28). A process must be brought into use by the firm.

The definitions of innovation in the third and fourth editions of the *Oslo Manual* are similar (discussed in Chapter 6). This has the advantage that statistics gathered through surveys based on the *Oslo Manual* require little additional guidance of respondents. The example is the Community Innovation Survey (CIS) conducted by Eurostat every two years, the history of which, up to 2012, is reviewed by Arundel and Smith (2013). CIS questions are revised regularly, and modules are added to study topics of policy interest, while the definition remains essentially the same.

7.3.2   Going beyond the Market

The fourth edition of the *Oslo Manual* provides a general definition of innovation, applicable to all economic sectors, and the basic difference from previous definitions of business innovation is that the product innovation has to be 'made available to potential users', which is broader than 'introduced on the market'.

One of the ways of making a product available to potential users is to put it on the market at economically significant prices and that has been part of the definition of product innovation for decades and surveys have collected data that have populated indicators which are part of official statistics. However, there are other ways of making a product available to potential users, such as providing it at a price of zero or any other non-economically significant price. Products that are new or improved and differ significantly from the firm's previous products which are not introduced to the market but are made available to potential users at zero price do not appear in official statistics on product innovation. This is an issue, described briefly in Chapter 6, now that the digital economy is so prominent.

With the rise of the digital economy, 'potential users' are being offered, at zero price, new or improved products such as apps and updates to their apps, access to cloud storage or computing, software, hubs and social media. Other products, such as the Internet of Things, may be provided at economically significant prices. A characteristic of many of these products, in addition to those characteristics that make it an innovation, is that they come with a link between the producer and the user that allows the use of the product to be monitored and the information used for various purposes. While zero price products are increasingly shaping the way people think, work and interact with society, they can also be instruments of cybercrime.[5]

There is at least as much justification for providing official statistics on zero price product innovations as for product innovation offered at economically significant prices. Statistics on both zero price and market price products should be available to policy developers and for monitoring and evaluation of existing policy interventions, leading to policy learning. There is an ongoing discussion of the products in the digital economy and their influence on the SNA (Diewert et al. 2018; Brynjolfsson et al. 2019).

7.4   GENERAL GOVERNMENT SECTOR AND PUBLIC SECTOR

General government consists of institutional units that, in addition to fulfilling their political responsibilities and their role of economic regulation, produce services (and possibly goods) for individual or collective consumption mainly on a non-market basis and redistribute income and wealth (EC et al. 2009: para. 2.17c).

The general government sector and the public sector are described in more detail in Chapter 22 of the *2008 SNA* (EC et al. 2009: 435). The chapter starts with general government units which include some non-profit institutions (NPIs) and public enterprises not treated as corporations. The public sector includes general government and public corporations.

These are the institutional units which are the target for surveys of innovation in the general government sector and the public sector. Chapter 22 discusses how to identify the institutional units that belong in the general government and the public sector. Once the institutional units are identified a survey can follow, using the general definition of innovation to identify innovative units.

There is no manual on innovation in the general government sector or the public sector, although this has been discussed at the OECD over the years. The empirical work on public sector innovation is reviewed in Gault (2012, 2015, 2016) and further discussed in the context of a general definition in Gault (2018a).

An early study of public sector innovation was the 2011 project for Measuring Public sector Innovation in the Nordic countries (MEPIN).[6] The objective of the project was to develop a measurement framework for collecting internationally comparable data on innovation in the public sector in order to understand what public sector innovation is and how public sector organisations innovate. The measurement was expected to lead to indicators that could support the promotion of public sector innovation (Bloch and Bugge 2013). This is discussed in Gault (2015, 2018a).

To study public sector innovation, the OECD has created the Observatory of Public Sector Innovation (OPSI) which collects examples of innovation but does not propose a definition of public sector innovation. It does provide a platform to support a network of practitioners[7] and gathers information on case studies and work on public sector innovation. In addition, the EU has presented survey results and related information in the European Public Sector Innovation Scoreboard 2013 (EPSIS) (EC 2014).

Australia has been active in measuring public sector innovation and Arundel and Huber (2013: 146) have noted the absence of an agreed definition and used 'public sector innovation involves novelty and the intention of making something better, for instance through new or improved services or processes'. This definition of innovation raises questions about time scale and the meaning of 'better'. The authors have suggested that public sector managers think of innovation as improving the current state of the unit and they see 'innovation' as an ongoing process. This differs from the discussion of the definitions in Chapter 6 and raises a question about surveys, their development, cognitive testing and their use. Do you work with what the respondent thinks is the objective of the study or do you apply a standard definition, in this case of innovation, and then guide the respondent?

Ideally, an innovation survey should not use the word 'innovation'. Using the general definition, the first innovation question, as discussed, is about new or improved products or processes. The second, or it may be combined with the first, is about whether there is a significant difference from the unit's previous products or processes. The third question is about what was done with the product (made available to potential users?) or process (brought into use by the institutional unit?). Nowhere should the respondent be asked about 'innovation'. The survey instrument can be used as a means of teaching respondents about how to respond to the questions and once that is done, the statistician can infer the presence of innovation, as defined in the *Oslo Manual*, or not.[8]

From the perspective of a manual for collecting and interpreting data from the public sector, Arundel et al. (2019: 789) have suggested that 'there is sufficient evidence, drawn from surveys of innovation in the public sector and cognitive testing interviews with public sector managers, to develop a framework for measuring public sector innovation'. This would bring together considerable experience from the MEPIN project and work in Australia (Arundel et al. 2015) to make a case that could be put to the OECD and the EU for the development of a manual. In the manual, the general definition of innovation could be applied to the public sector.

Manuals for statistical measurement of innovation, in different sectors, raise some questions. Since 1992, the OECD Working Party of National

Experts on Science and Technology Indicators (NESTI) has been responsible for the guideline on the measurement of innovation in the business sector. From 1997, the relevant section of Eurostat became jointly responsible for the development and publication of the *Oslo Manual*. If there is to be a manual for the collection of information on innovation in the public sector, the household sector and the NPISH sector, there will have to be close cooperation with the relevant subject matter sections in the OECD and Eurostat. This is a potential challenge for the development of the subject of innovation measurement and related policy development.

7.5   NON-PROFIT INSTITUTIONS SERVING HOUSEHOLDS

NPISHs are legal entities that are principally engaged in the production of non-market services for households or the community at large and whose main resources are voluntary contributions (EC et al. 2009: para. 2.17e). NPISHs are not controlled by government.

Examples of NPISHs are professional or learned societies, political parties, trade unions, consumers' associations, churches or religious societies, and social, cultural, recreational or sports clubs (EC et al. 2009: para. 4.167). There is no manual for NPISH sector innovation, but the general definition of innovation can be applied.

7.6   HOUSHOLD SECTOR

Households are institutional units consisting of one individual or a group of individuals in a household. All physical persons in the economy must belong to one and only one household. The principal functions of households are to supply labour, to undertake final consumption and, as entrepreneurs, to produce market goods and non-financial (and possibly financial) services. The entrepreneurial activities of a household consist of unincorporated enterprises that remain within the household except under certain specific conditions (EC et al. 2009: para. 2.17d, para. 4.156). There is no manual for household sector innovation similar to the *Oslo Manual* for the business sector. As with other sectors without manuals on measuring innovation, the general definition can be applied.

Members of households may develop product or process innovations, but for them to be innovations they must be able to confirm that there is a *new or improved product or process (or combination thereof) that*

differs significantly from the household's previous products or processes.
They must then be able say that the product had been made available to
potential users or that the process had been brought into use by the household's unincorporated enterprise.

The challenge in dealing with innovation in the household sector is
classifying the household activity. If the activity is to run an unincorporated enterprise within a household, then it can be dealt with in the same
way as the firms in the business sector; a product or process is identified
satisfying the criteria of the general definition in the fourth edition of the
Oslo Manual and the product has been made available to potential users
or the process has been brought into use by the unincorporated enterprise.
Larger unincorporated enterprises, such as law offices, can be treated as
quasi-corporations in the business sector.

It is also possible for a household, including individuals in the household, to innovate by changing products acquired as consumers to enhance
the benefit derived from the improved product. For there to be innovation,
the product must be improved, and it must be made available to potential
users.[9] The household could also create a product which was new if the
product could not be found on the market. As with the improved product,
it must be made available to potential users to be an innovation.

Work has been done on consumer innovation in households (von
Hippel 1988, 2005, 2007, 2016, 2017; Harhoff and Lakhani 2016) where
the focus has been on 'new or improved products' with less emphasis on
the action of making them available to potential users. However, there is
an active community engaged in this subject and it was during a study
of user innovation for the government of Finland (de Jong et al. 2015)
that the action making the product available to potential users was first
discussed (Gault 2012). There were three means of making the product
available to potential users: providing a prototype, or the knowledge
needed to produce the prototype, to a firm that produced the product;
starting a business to provide the product, or adding a new line of business to an existing business; or making the product available to potential
users such as a peer group or a community of practice. The first two could
be found in official innovation statistics, but not the third.

More needs to be done on household innovation which involves
making the product available to peer groups or communities of practice.
While there is a long history of studying household sector innovation
(von Hippel 2017) the subject has yet to be brought into official statistics.

7.7 THE REST OF THE WORLD

The sectors divide up the economic activity in the country, however, countries have borders and tangible and intangible things cross them in both directions. The Rest of the World (ROW) is discussed in Chapter 26, 'The rest of the world accounts and links to the balance of payments' of the *2008 SNA*. Where the ROW and the balance of payments statistics enter the innovation domain is in the linkages that connect institutional units in the country to those outside of the country. This is relevant to the study of global value chains and supply chains and their place in the study of innovation.

7.8 CONCLUSION

This chapter has introduced the SNA and the economic sectors of the SNA and it has explored the measurement of innovation, as defined in Chapter 6, in all economic sectors. This has raised questions about the absence of official statistics on innovation in all but the business sector and the impact on innovation statistics if innovation were measured everywhere. In the business sector, product innovations made available at zero price or, more generally, at non-economically significant prices, are not present in official statistics which is a significant gap in understanding innovation, especially in a digital economy.

The focus in this chapter has been on innovation in each economic sector. However, there are types of innovation that can happen in any economic sector and that is the subject of the next chapter.

NOTES

1. This may seem a more detailed discussion than is needed. However, there are researchers that dismiss innovation surveys as opinion surveys, not as statistical measurement of the activity of innovation. For the two components of the definition, the respondent will have an opinion about whether there is a new or improved product or process. The respondent should be able to distinguish whether the new or improved product or process differs significantly from previous products or processes, or not, and know whether the product has been made available to potential users or the process brought into use.
2. For readers not familiar with the SNA and GDP, consider Coyle (2016).
3. This is a decision needed to deal with an inconsistent use of households sector in the *2008 SNA*. Households sector is used in Chapters 24 and 25 and

household sector in most other chapters with some exceptions of which EC et al. (2009: para. 4.24) is one.

4. There was an attempt to capitalise R&D in the SNA 1993 revision which failed. There were many considerations, one of which was that GDP would go up once R&D was a capital expenditure and that would increase the cost of membership in some international organisations. Official statisticians noted that if R&D were capitalised, it would be part of capital expenditure statistics which were an important part of the SNA. Innovation statistics are not part of the SNA.

5. Like innovation, cybercrime is everywhere. In the UK, it is taken very seriously. See https://www.nationalcrimeagency.gov.uk/what-we-do/crime -threats/cyber-crime (accessed 17 March 2020).

6. See http://nyskopunarvefur.is/files/filepicker/9/201102_mepin_report_web .pdf (accessed 17 March 2020).

7. See https://oecd-opsi.org/ (accessed 17 March 2020).

8. In addition to using the survey instrument as a means of guiding the respondent, it can also be used to teach the respondent about the subject being studied. See Gault (2013: 17).

9. 'Potential users' are just that, people or households that might benefit from using the product. 'Potential users' excludes people who are not likely to use the product.

8. Measuring innovation across economic sectors

8.1 INTRODUCTION

In Part III, Chapter 6 introduced the general definition of innovation, applicable in all economic sectors, and Chapter 7 provided the definitions of the economic sectors in the System of National Accounts (SNA). This led to a discussion of sector-specific innovation. In Chapter 8, the goal is to examine innovation that can occur in any of the SNA sectors. This raises some problems, but it leads to measurement challenges for future work, including how to develop policy for innovation that can happen everywhere.

In the examples which follow, the general definition is applicable. They are: innovation in the informal economy; ecological innovation (eco-innovation); and innovation involving general purpose technologies and practices. The general purpose technology section includes a discussion of user innovation in any sector.

While the three examples of innovation can be specified by the general definition, there are cases where this does not work. An example is 'social innovation'. It is policy relevant and the subject of an extensive policy and research literature, but the approaches to the subject are different and the definitions of social innovation diverse. This will be discussed briefly along with the difference between definitions that support discussion and theorising, introduced in Chapter 6 (Section 6.3), and the general definition which supports the measurement of innovation and the review of innovation policy.

8.2 INNOVATION IN THE INFORMAL ECONOMY

The informal economy (or sector) is present in every country and more so in developing countries. The term 'sector' does not correspond to sectors used in the SNA to cover all of the economy sectors (EC et al. 2009: para. 25.37). The informal aspects of the economy are discussed in Chapter 25 of the SNA. Chapter 25 also includes the International Labour Organization (ILO) definition of the informal sector from the 15th International Conference of Labour Statisticians (ICLS); the 'informal sector' is a subset of household unincorporated enterprises (EC et al. 2009, para. 25.36).

In addition to understanding informal production, the distinction between formal and informal employment is also relevant (EC et al. 2009: para. 25.54) and plays a key role in the statistical measurement of the informal sector. Charmes (2016, 2019) describes the use of employment surveys to identify informal units. A survey of the units may then follow to gather information on informal activities.

The interest here is in the statistical measurement of innovation in the informal sector. Charmes (2019) has reviewed the last 50 years of conceptualisation and measurement of the informal sector. Survey methods are also reviewed in Charmes et al. (2018). Charmes (2016) and Charmes et al. (2016) discuss how and where questions on innovation could be inserted into existing economic surveys of households and firms. One of the approaches is to consider to what extent questions from the EU Community Innovation Survey (CIS) can be used in surveys of innovation in the informal economy and what new questions could be added and others removed (Charmes 2016; Charmes et al. 2016). The advantage of this approach is that the statistical measurement of the informal sector, and innovation in the (formal) business sector, are well established with well-tested questions, questionnaires and survey methodology. Bringing both together, while removing questions which do not fit is a first step to a comprehensive measurement programme. The next step is to use the data resulting from surveys to produce indicators that can support policy development.

A new step, made possible by the general definition of innovation, is to replace 'introduced on the market' from the business sector definition of innovation by 'made available to potential users'. Once this is done, questions can be posed about product innovations that are made available at non-economically significant prices, including zero prices. This is important as actors in the informal sector may transfer product innovations for cultural, religious or social reasons that have nothing to do with the market. The general definition of innovation makes possible the gathering of data, providing indicators for monitoring and evaluating innovation policy in the informal economy.

8.3 ECO-INNOVATION

While countries look to innovation to provide jobs and economic growth, there is interest not just in innovation but in eco-innovation leading to a green economy. An example is the European Union green.eu project and the *Maastricht Manual on Measuring Eco-Innovation for a Green Economy* (Kemp et al. 2019). The project recognises the urgency of having robust statistical measurement of eco-innovation to support international comparisons as countries move towards green economies. The *Maastricht Manual* goes beyond a standard statistical manual to discuss policies for eco-innovation and the green economy. This includes the measurement of the effect of implemented policies and the monitoring and evaluation of the policies.

Eco-innovation is a means of moving towards a greener economy, making the world a better place, but the development of effective policy that leads to this requires robust and internationally comparable statistical indicators to guide policy development and to monitor and evaluate policies that have been implemented. To achieve this, there must be guidelines for the statistical measurement of eco-innovation. The *Maastricht Manual* is a significant step towards having official statistics on eco-innovation that can support research on eco-innovation.

The *Maastricht Manual* encourages the measurement community, statistical offices and research institutes, to add statistical measurements of eco-innovations, and their outcomes, to their official statistics. Official statistics are important because they are credible and support informed public discourse on environmental priorities and the allocation of resources to implementing more effective green policies. In addition to setting standards for data collection and interpretation the *Maastricht Manual* educates the reader, encourages the use of the data and indicators and should help to build communities of practice that are trying to contribute to the green economy.

This manual is not an end but a beginning. The OECD Working Party of National Experts on Science and Technology Indicators (NESTI) spent many years in discussion before its collective knowledge of innovation was first codified in the *Oslo Manual* of 1992. Once the manual existed, statisticians, researchers, policy analysts and policy makers learned from collecting and using statistics on innovation and applied that knowledge to revise the manual which is now in its fourth edition (OECD/Eurostat 2018).

Eco-innovation can happen everywhere, in the public sector, the government sector, the business sector, the private non-profit sector serving households and the household sector. In the past, the interest has been focused on the business sector, but now the *Oslo Manual* provides a general definition of innovation applicable in all sectors. This applies, as well, to eco-innovation. The definition of eco-innovation is the following:

> An **eco-innovation** is a new or improved product or practice of a unit that generates lower environmental impacts, compared to the unit's previous products or practices, and has been made available to potential users (product) or brought into use by the unit (process). (Kemp et al. 2019: Chapter 2)

The definition is close to that of the general definition in the fourth edition of the *Oslo Manual* (OECD/Eurostat 2018: para. 2.99) and it is restricted (see Chapter 6) by the addition of 'that generates lower environmental impacts'. The closeness of the definition of eco-innovation to the general definition of innovation provides a link to the broader innovation community and NESTI. It also raises a question about the institutional home for the eco-innovation community as this is where participants can share knowledge gained by doing surveys and analysing the resulting data. In time, this could provide the basis for the next revision of the manual.

The *Maastricht Manual* is a new tool for studying eco-innovation, for learning about this activity, and supporting policy that can lead to a greener economy.

8.4   TECHNOLOGIES, PRACTICES AND INNOVATION

This section examines general purpose technologies and their relationship to innovation in any economic sector.

8.4.1   General Purpose Technologies and Practices

A general purpose technology is generic and can be used in many applications.[1] Such technologies have appeared in industrial revolutions in the past: the steam engine; electricity; mass production; and ICT.[2] There is now a suggestion that the fourth industrial revolution (Schwab 2017, 2018) is happening at a time when the world is threatened by climate change, economic and social inequality, and challenged by the Sustainable Development Goals.

General purpose technologies of today, such as digitalisation, artificial intelligence (AI), robotics, the internet of things and the use of big data are moving quickly. The interest 20 years ago was in the access to the internet, the web, and the use of the web to inform clients of available products. Later, the questions concerned e-commerce on the web, both buying and selling. A significant difference is that 20 years ago everything that took place on the web was programmed, by people, whereas now AI is used to find patterns in (big) data, to make decisions and to revise code in apps. Technologies are not just goods related, they can include services and practices such as business models, knowledge management and the use of value and supply chains.

There is also a change in how innovation is managed in the business sector (OECD 2019a). Previously product innovation took place when a new or improved product was introduced on the market, at market prices. Now there are new or improved products offered for zero price, discussed in Chapter 6, Section 6.3. They are still product innovations, but the difference is that these zero price products, in many cases, connect the producer and the user, allowing information about the user to be captured by the producer. Examples are apps, email accounts, access to the iCloud and electronic hubs for trading or other activities. This is a radical change in the business sector, and it can happen in all of the other SNA sectors where institutional units can be users or producers of product innovations. The flow of information from users to producers can be used to make potential users aware of similar products or to suggest alternatives. Such information can also be misused.

So far, only technologies have been discussed, but the same discussion can apply to services and practices, such as knowledge management practices (Foray and Gault 2003) or organisational practices.

General purpose technologies can also be combined to do new things, and this is evident in the digital economy. These activities are not limited to the business sector; they can be used to innovate in any sector. Technologies, general purpose, or more specific, can give rise to process innovation, with implications for user innovation.

8.4.2   Process Innovation and User Innovation

From a measurement perspective there are two things to consider when dealing with process innovation. The first is that process innovation can occur in an institutional unit if the unit acquires, and brings into use, a product that is, from the perspective of the unit, new or improved and differs significantly from the unit's previous processes. This process innovation draws upon products first developed by other institutional units, with little or no additional modification (OECD/Eurostat 2018: para. 3.14, generalised). In the business sector, this process innovation by acquisition (purchase, rental or leasing) of 'off-the-shelf technologies' is distinguished from user innovation (Gault 2016).

'User innovation' occurs when the institutional unit acquires a product for use and then modifies it to serve better the purposes of the institutional unit. A second type of user innovation occurs when the institutional unit cannot find the technology, or practice, that is needed, and it adopts the technology or practice by developing it for its own use (Gault 2016).

8.4.3 Policy

This is a measurement chapter but the use of general purpose technology can influence policy. There are three levels of innovation policy to consider. The first is the acquisition of products that are new or improved to the institutional unit and which allow the institutional unit to be classified as a process innovator. The acquisition can range from procurement policy in public institutions to acquisition by small household unincorporated enterprises. Policy will vary from sector to sector, but the policy maker may wish to promote the acquisition and use of general purpose technologies, such as broadband web access to support electronic commerce anywhere in the country, assuming the infrastructure is in place to support it.

As institutional units will acquire products that are readily available and are new or improved from the perspective of the institutional unit, any statistical measurement would expect to show that a high percentage of institutional units qualify as innovative, based on their acquisitions.

As already discussed, institutional units can acquire products and modify them to meet their own requirements better, or they can develop such products for their own use if they are not readily available. These are cases of user innovation and policy can support this and other types of innovation by promoting the training of workers, by encouraging links to education or research institutions to make use of the knowledge and experience that can be found there, or to encourage collaboration by institutional units.

8.5   SOCIAL INNOVATION

Social innovation is a subject of policy interest and it covers a wide range of topics, including social value[3] and ethics. There are no comparable official statistics based on an internationally accepted definition.[4] The definition of innovation for measurement purposes in the fourth edition of the *Oslo Manual* is not normative and it does not address value or ethics unless these are imposed as a restriction (Chapter 6, Section 6.3).

An example of the range of definitions is provided by Balamatsias (in 2018) in the Social Innovation Blog of the Social Innovation Academy where the 'eight most popular definitions' are listed.[5] None of the eight definitions are designed to support statistical measurement and all include expected outcomes as part of the definition. The European Commission, as part of its innovation policies, provides a definition of social innovation.[6] Social innovations are:

> new ideas that meet social needs, create social relationships and form new collaborations. These innovations can be products, services or models addressing unmet needs more effectively.

The difficulty of defining social innovation is illustrated by Moulaert et al. (2013: 15), Moulaert and MacCallum (2019), Nicholls and Murdock (2012) and by Rüede and Lurtz (2012).[7]

Mulgan (2007:8) provides the following definition. Social innovation comprises:

> innovative activities and services that are motivated by the goal of meeting a social need and that are predominantly developed and diffused through organisations whose primary purposes are social.

As an example of applying this definition, consider the 'Delayed transfers of care' statistics in the UK (House of Commons Library Briefing Paper 2019). The issue is that patients are in the wrong care setting which can arise when care settings managed by the social services, or the local government, lack the capacity to transfer patients from hospitals to social service care settings. The policy intervention could be an integration of social services and hospital care services to support planning and resource allocation with a view to moving more people to the right care setting at the right time.

The implementation of the policy could involve new or improved services made available to potential users, those responsible for the care facilities, and a new or improved process involving management and planning brought into use by the institutional units, in this case the hospital and the social services. There is no guarantee that the desired improvement will happen, but its presence can be confirmed by observation (surveys or administrative data) and change can be monitored over time and comparisons made with other jurisdictions where the policy has been implemented and those where it has not.

This example is chosen as it could be used to support measurement but with a definition of 'innovation', qualified by the word 'social', and of 'innovative activities' which could correspond to innovation activities in the fourth edition of the *Oslo Manual* (OECD/Eurostat 2018: 247). The definitions diverge when an expectation is added 'meeting a social challenge' as well as the means of diffusion, 'through organizations whose primary purposes are social'. This suggests that the definition of Mulgan et al. (2007) fits into the category introduced in Chapter 6 (Section 6.3), as a definition 'to guide discussion and theorising' which is an important part of developing the subject, in this case, social innovation. The question is whether the social innovation community sees a need for statistical measurement leading to monitoring and evaluation in social innovation policy in any economic sector.

Going back to business sector innovation, at least a decade was spent experimenting with surveys of technological innovation before the community could agree on a definition of innovation and how product or process innovation could be measured in the manufacturing sector (OECD 1992). It could be argued that social innovation is considerably more complex than innovation in manufacturing and it will take longer to build a consensus on measurement. Missing in this process is the equivalent of NESTI which, in consultation with Eurostat, oversees innovation manuals and their revision. In addition, the measurement of the activity of social innovation may not be encouraged. Murray et al. (2010: 6) note that 'Measuring success in the social economy is particularly problematic' and then go on to explain why. Another approach is taken by Marée and Mertens (2012) when discussing the measurement of the performance of social innovation.

The importance of statistical measurement of social innovation, as with the other types of innovation discussed in this chapter, is that the resulting indicators support policy development and then provide a basis for monitoring the implemented policy. Not discussed in this chapter, but also important, is the way the growing digital economy is influencing innovation of all kinds, including social innovation. This is discussed in Part IV.

8.6   CONCLUSION

This chapter has examined challenges of measuring innovation in the informal economy, eco-innovation, and innovation driven by technologies and practices, both generic and more specific. The general definition of innovation makes it easier to measure the presence of innovation in any economic sector, whether or not the innovation is sector specific (e.g., public sector innovation) or potentially present in all sectors (e.g., green innovation). Social innovation is presented as a work in progress where the general definition does not yet apply, and it may be some time before there are robust measures of social innovation showing change over time and providing a basis for comparison across countries.[8]

Chapters 6, 7 and 8 have examined the measurement of innovation in all sectors of the economy, either sector specific or occurring in any sector. The next task is to see how the general definition can be applied to global challenges, how measurement challenges can be overcome, and how both relate to policies.

NOTES

1. See Lipsey et al. (2005) on general purpose technology and its application.
2. See https://ec.europa.eu/eurostat/cache/metadata/en/isoc_se_esms.htm (accessed 17 March 2020) and note that the ICT sector includes goods producing industries and services producing industries.
3. See Mulgan et al. (2013) and Mulgan (2019) on measuring social value.
4. See Moulaert et al. (2013) and Howaldt et al. (2014) for an overview of the subject and a critical view of the literature.
5. http://www.socialinnovationacademy.eu/8-popular-social-innovation -definitions/ (accessed 17 March 2020).
6. See https://ec.europa.eu/growth/industry/innovation/policy/social_en (accessed 17 March 2020).
7. Other work on social innovation can be found at: https://www.siceurope.eu/ about-sic/what-socil-innovation/what-social-innovation (accessed 17 March 2020).
8. International comparison may not be an objective of all social innovation communities. The author attended a parallel session of participants in a social innovation conference and asked about establishing baselines and monitoring policy intervention. He was told that, as he was clearly not committed to the cause, he would find another session more to his liking.

PART IV

Where next?

9. Innovation and future challenges

9.1   INTRODUCTION

This chapter draws upon the earlier chapters of this book to comment on innovation measurement and how it might be of use in monitoring or evaluating policies. What is different is that the innovation discussed here may be happening in any or all of the economic sectors, not just the business sector, and the activities in the sectors may be linked. This emphasizes the systems dimension as part of understanding the role of innovation and the monitoring and evaluation of innovation policy.

The world is changing rapidly, and, for some time, there has been policy interest in sustainable development and climate change, topics which overlap. While the environment is changing in response to human activity, the economy is changing as it becomes progressively more digital. The challenge is understanding these changes, through statistical measurement, leading to indicators that inform policy development, and not just measurement of activities in economic sectors, in isolation.

In 2015 the 17 Sustainable Development Goals (SDGs) (Box 9.1) were introduced with the expectation that the targets would be met by 2030 (United Nations 2015). The task in this chapter, indeed in this book, is to examine how understanding innovation, in any or all economic sectors, can help. This is not as simple as it might seem.

Turning to climate change, there is a Framework for the Development of Environmental Statistics (FDES) but there is no reference to 'innovation' in the indicators present in the framework. As with the SDGs, the absence of 'innovation' is an issue considered here.

'Green growth' and the expectation that green activities and innovation protect the environment while generating economic growth are considered, as are innovation policies for inclusive growth. In both cases, the importance of the innovation systems approach and analysis of innovation is emphasized.

International health issues are considered and the capacity for innovation in a number of economic sectors to deal with the threats to health. This includes threats to cyber security in the health care system.

The chapter concludes with future directions for innovation measurement, policy monitoring and evaluation leading to policy learning.

9.2   SUSTAINABLE DEVELOPMENT GOALS AND MEASUREMENT

The definition of 'sustainable development' used in the SDGs is the following:[1]

> Sustainable development is development that meets the needs of the present without compromising the ability of future generations to meet their own needs.

BOX 9.1 SUSTAINABLE DEVELOPMENT GOALS

A.   People

Goal 1. End poverty in all its forms everywhere

Goal 2. End hunger, achieve food security and improved nutrition and promote sustainable agriculture

Goal 3. Ensure healthy lives and promote well-being for all at all ages

Goal 4. Ensure inclusive and equitable quality education and promote lifelong learning opportunities for all

Goal 5. Achieve gender equality and empower all women and girls

B.   Planet

Goal 6. Ensure availability and sustainable management of water and sanitation for all

Goal 12. Ensure sustainable consumption and production patterns

Goal 13. Take urgent action to combat climate change and its impacts*

Goal 14. Conserve and sustainably use the oceans, seas and marine resources for sustainable development

Goal 15. Protect, restore and promote sustainable use of terrestrial ecosystems, sustainably manage forests, combat desertification, and halt and reverse land degradation and halt biodiversity loss

C.   Prosperity

Goal 7. Ensure access to affordable, reliable, sustainable and modern energy for all

Goal 8. Promote sustained, inclusive and sustainable economic growth, full and productive employment and decent work for all

Goal 9. Build resilient infrastructure, promote inclusive and sustainable industrialization and foster innovation

Goal 10. Reduce inequality within and among countries

Goal 11. Make cities and human settlements inclusive, safe, resilient and sustainable

D.   Peace

Goal 16. Promote peaceful and inclusive societies for sustainable development, provide access to justice for all and build effective, accountable and inclusive institutions at all levels

E.   Partnership

Goal 17. Strengthen the means of implementation and revitalize the Global Partnership for Sustainable Development

Note: * United Nations (2015) acknowledges that the United Nations Framework Convention on Climate Change is the primary international, intergovernmental forum for negotiating the global response to climate change.

Source: United Nations (2015:14).

Looking at the SDGs in Box 9.1, it is clear that there is no substantive use of the word 'innovation'. It appears once, in the title of SDG 9; innovation is to be 'fostered'. Each SDG has a number of sub-goals, for a total of 169 (United Nations 2015), and for each sub-goal there are indicators, of which 232 are unique. There are six occurrences of 'innovation' in the 169 sub-goals and none in the 232 indicators (United Nations 2019a). What is the role of innovation in the SDGs?

The SDGs (Box 9.1) provide (non-obligatory) targets for the year 2030 for participating countries (United Nations 2015). There are 17 goals with the 17th being the implementation of the other 16. In Box 9.1 the SDGs are divided into five categories:[2] People (1–5); Planet (6, 12–15); Prosperity (7–11); Peace (16); and implementation of the 17 goals (17).

The SDGs are seen by some as targets for developing countries, as were the Millennium Development Goals (MDGs), however, in the UK, 30 per cent of children were in poverty in 2017/18 after housing costs were taken into account (Francis-Divine et al. 2019: 12), and in the EU, 'Almost every fourth person in the EU is still at risk of poverty or social exclusion …' (Eurostat 2018). Poverty, SDG 1, is a global issue as are the other SDGs.

As the focus of this book is on the measurement of innovation and the use of the resulting indicators to support government policy, business strategy or household planning, the five categories of the SDGs are examined from an innovation measurement perspective. This is not a comprehensive analysis, but an indicative one suggesting, based on the sub-goals that mention innovation, areas in which statistical measures of innovation could contribute to the monitoring of selected SDGs as new indicators are developed in later stages of the SDG process.

9.2.1   People

There is no reference to innovation in goals 1–5, but there is scope for innovation indicators that show new or improved products or processes that are made available to potential users or brought into use in the institutional unit and have outcomes that change the state of poverty, ideally reducing it. The institutional units could be in any economic sector, including government departments, education institutions, hospitals, health care services, businesses, households and non-profit organisations serving households.

9.2.2 Planet

Goals 6, 12–16 address issues affecting the planet. There is no reference to innovation in the goals, the sub-goals or the indicators.

9.2.3 Prosperity

Goals 7–11 deal with the economy. In sub-goals 8.2 and 8.3 there is reference to innovation. Sub-goal 8.2 is to: 'Achieve higher levels of economic productivity through diversification, technological upgrading and innovation, including through a focus on high-value added and labour-intensive sector'. The indicator is 8.2.1, 'The annual growth rate of real GDP per employed person'.

Sub-goal 8.3 is to: 'Promote development-oriented policies that support productive activities, decent job creation, entrepreneurship, creativity and innovation, and encourage the formalisation and growth of micro-, small- and medium-sized enterprises, including through access to financial services'. The indicator is 8.3.1, 'Proportion of informal employment in non-agriculture employment, by sex'.

In both sub-goals 8.2 and 8.3, innovation is promoted but there is no guidance on measuring innovation and its impact on the institutional units that are involved in achieving the sub-goals as part of achieving SDG 8.

In SDG 9, an objective is to 'foster innovation'. This arises in two sub-goals, 9.5 and 9.b.

Sub-goal 9.5 is to: 'Enhance scientific research, upgrade the technological capabilities of industry sectors in all countries, in particular developing countries, including, by 2030, encouraging innovation and substantially increasing the number of research and development workers per 1 million people and public and private research and development spending'. There are two indicators, 9.5.1, 'Research and development expenditure as a proportion of GDP'; and 9.5.2, 'Researchers (full-time equivalents) per million inhabitants'. The indicators deal with the allocation of resources, financial and human, to the performance of R&D. However, R&D is not innovation.

Sub-goal 9.b is to 'Support domestic technology development, research and innovation in developing countries, including by ensuring a conducive policy environment for, inter alia, industrial diversification and value addition to commodities'. The indicator is 9.b.1, 'Proportion of medium and high-tech industry value added in total value added'. As with Goal 8, 'innovation' is not present in the indicator.

9.2.4   Peace

There is no reference to innovation in Goal 16, but there are opportunities to develop policies on public sector innovation and indicators that could be used once the policies had been implemented.

9.2.5   Partnership

Goal 17 deals with implementation of the other 16 goals. Innovation appears in two sub-goals, 17.6 and 17.8, under the heading of technology.

Sub-goal 17.6 is to 'Enhance North-South, South-South and triangular regional and international cooperation on and access to science, technology and innovation and enhance knowledge-sharing on mutually agreed terms, including through improved coordination among existing mechanisms, in particular at the United Nations level, and through a global technology facilitation mechanism'. There are two indicators, 17.6.1 and 17.6.2, but neither deals with innovation.

Sub-goal 17.8 is to 'Fully operationalize the technology bank and science, technology and innovation capacity-building mechanism for least developed countries by 2017 and enhance the use of enabling technology, in particular information and communications technology'. There is one indicator, 17.8.1, 'Proportion of individuals using the Internet'.

The 18 sub-goals of Goal 17 deal with finance, technology, capacity building, trade and systemic issues. Each of these topics is an area for public sector innovation, links to international organisations, and business sector innovation, linked to trade rules, and policy coherence.

In summary, 'innovation' appears in the text of six sub-goals: 8.2, 8.3, 9.5, 9.b, 17.6 and 17.8. It does not appear in any of the indicators. The observation is made in OECD (2019c: 19) that the 2030 Agenda was 'politically driven rather than based on a conceptual framework' and the 169 targets are heterogeneous. The issue here is not the heterogeneity but the understanding that could result from adding indicators, based on the direct measurement of innovation, related to sustainable development. That is an opportunity for future work on sustainable development and the role of innovation and the development of policy in any economic sector that supports the SDGs. This also fits with the work of the Global Partnership for Sustainable Development.[3]

9.3   CLIMATE CHANGE

9.3.1   The UN Framework Convention on Climate Change

Climate change has been an issue for decades and more formally after the Rio Earth Summit in 1992. The UN Framework Convention for Climate Change (UNFCCC) started there and was released on 21 March 1994. The Kyoto Protocol, linked to the UNFCCC, came into being on 16 February 2005, ending in 2012. The Paris Agreement builds on the UNFCCC and became effective on 4 November 2016.

Measurement activities related to climate change have focused on the temperature of the planet, greenhouse gas emissions and gases damaging the ozone layer. The Paris Agreement includes proposals for mitigation of climate change[4] which could be part of an innovation policy in a participating country, but the term 'innovation' is not prominent.

There is work by the UN Statistics Division[5] on 'climate change statistics and indicators and adaptation-related SDG indicators' where 'adaptation' describes, with 'mitigation', the human response to climate change. The following SDG indicators in Table 9.1 are identified and related to the corresponding statistics from the FDES. The point has already been made that innovation is not prominent in indicators used to describe the SDGs. In the FDES (United Nations 2017), 'innovation' appears once, in a quotation of OECD work on green growth (para. B. 23).

As with the SDGs, there is knowledge to be gained from the direct measurement of innovation related to climate change, in any economic sector.

9.4   GREEN GROWTH AND ECO-INNOVATION

In 2009, 34 Member Countries of the OECD provided a mandate to the OECD to develop a Green Growth Strategy which was delivered in 2011. A definition was provided: 'Green growth means fostering economic growth and development, while ensuring that natural assets continue to provide the resources and environmental services on which our well-being relies. To do this, it must catalyse investment and innovation which will underpin sustained growth and give rise to new economic opportunities' (OECD 2011: 4).

Table 9.1   SDG sub-goals related to the corresponding FDES statistics

SDG sub-goal	Content
1.5.3/11.b.2/13.1.2	Number of countries that adopt and implement disaster risk reduction strategies
4.7.1(ii)	Extent to which education for sustainable development is mainstreamed at all levels
6.4.1	Change in water use efficiency over time
7.2.1	Renewable share in the total final energy consumption
11.4.1	Total expenditure per capita on preservation, protection and conservation of all cultural and natural heritage
11.c.1	Proportion of financial support to the least developed countries that is allocated to the construction and retrofitting of sustainable, resilient, and resource-efficient buildings using local materials
12.5.1	National recycling rate, tons of material recycled
12.a.1	Amount of support to developing countries on research and development for sustainable consumption and production and environmentally sound technologies
12.c.1	Amount of fossil-fuel subsidies per unit GDP and as a proportion of total national expenditure on fossil fuels
13.3.1	Number of countries that have integrated mitigation, adaptation, impact reduction and early warning into primary, secondary and tertiary curricula
14.5.1	Coverage of protected areas in relation to marine areas
15.3.1	Proportion of land that is degraded over total land area

Source: UN Statistics Division (2018, 2019a, 2019b).

The European Commission promotes green growth and the circular economy,[6] and, in a separate project, a team has developed the *Maastricht Manual on Measuring Eco-Innovation, Green Growth and the Green Economy* (Kemp et al. 2019).

Both the OECD and the EC activities provide a basis for innovation policy implementation and statistical measurement in all economic sectors. Governments can take direct initiatives or provide framework conditions that influence green growth or innovation. Firms can engage in product or process innovation that takes account of green growth or eco-innovation.[7] Indeed, they may take advantage of regulation to innovate and capture more of the market or avoid taxes on activities that are not green or ecologically sound.

With innovation happening and being measured in all economic sectors, there is an opportunity to identify and measure linkages between the actors in the different sectors.

The *Maastricht Manual* provides guidance on measuring eco-innovation and it will be a basis for an ongoing discourse on eco-innovation and how to measure it. The experience from the OECD is that, once there is a manual, Member Countries will learn from their measurement experience and seek to revise the manual, a matter that requires a consensus of OECD Member Countries. The team responsible for the *Maastricht Manual* may wish to seek an institutional home for the project so that there is a centre for discourse among experts, collective learning and revision of the manual.

9.5   INNOVATION AND INCLUSIVE GROWTH

While innovation is being promoted to address the issues just discussed, inequality has been growing in developed (OECD 2008, 2015b) and developing (Cozzens and Thakur 2014; Dutrénit and Sutz 2014) countries.

'Green growth' and 'innovation and inclusive growth' are expected to provide economic growth and additional outcomes; green activities and inclusion. These additional outcomes fit into the category of 'restricted' innovation (Chapters 5 and 6). However, the point is made by the OECD (2015b: 61) that inclusive innovation policies and their evaluation mechanisms are still experimental and can benefit from 'trial and error'. This echoes the history of the measurement of innovation in the business sector where over a decade was spent engaging in 'trial and error' before a consensus was reached on the definition of innovation and the first edition of the *Oslo Manual* (OECD 1992) was published.

Throughout this book, the point has been made that the statistical measurement of innovation produces indicators that can be used for monitoring and evaluation leading to policy learning. In the case of innovation for inclusive growth, this is a work in progress but there is a need for policy research, experimentation and monitoring, resulting in improved policy. There is also a need for international openness (OECD 2015d:61). This has been the role of NESTI for the development of the definition of innovation for use in the business sector and, in the fourth edition of the *Oslo Manual* (OECD/Eurostat 2018), for all economic sectors.

From a measurement perspective, so long as there is an agreed definition of 'inclusive innovation', data can be gathered from the economic sectors which contribute to inclusive innovation and used to produce indicators that support monitoring and evaluation of inclusive innovation policies.

Planes-Satorra and Paunov (2017) review the subject and provide examples of inclusive innovation. They start with a review of innovation policy and then go on to inclusive innovation policy, dividing it into three categories, social inclusiveness, industrial inclusiveness and territorial inclusiveness. These categories can be elaborated to provide a definition for statistical measurement of restricted innovation.

For inclusive innovation the authors identify ten activities:

* building capacities
* addressing discrimination and stereotypes
* providing incentive to invest in inclusive innovation
* facilitating access to finance
* providing business development support
* promoting networks
* improving access to talent by small businesses
* accessing global knowledge and technology
* maximising the potential of existing assets
* attracting innovative firms to peripheral regions.

Innovation policies are then discussed and examples in various countries provided. Inclusive innovation policy can involve government policy, business strategy and household practices, suggesting that the introduction of policy, strategy or practices could benefit from better information on links to all sectors and using the general definition of innovation.

9.6   HEALTH ISSUES

Innovation in all economic sectors can address health problems arising from diseases, lifestyle or cyber security failures.

Smallpox virus has been eradicated but the virus is still held in two laboratories, one in the US, one in Russia. In September 2019 there was an explosion in the Russian laboratory but there was no release of the virus.[8] The debate on whether to destroy the viruses held continues. As people refuse to have their children vaccinated, measles, which had almost been eliminated in developed countries, is returning. Ebola virus is active in Africa and could move to other continents. Malaria treatment is becoming difficult as strains of the parasite are becoming more drug resistant. There are other examples.

While viruses and parasites evolve, so does the use of cyber tools in health care, raising cyber security[9] issues (Ghafur et al. 2019).

Addressing these issues is a challenge for innovation in all economic sectors. There may be a case for coordination of innovation through linkages of the actors in different sectors to collaborate to arrive at the desired outcome.

9.7   CONCLUSION

In this chapter a selection of global challenges was reviewed to explore the use of innovation policy in participating countries to address the challenges.

In the Sustainable Development and the Climate Change goals, the use of innovation policy is very limited, leading to a suggestion that innovation policy be developed to address these global challenges and that the general definition of innovation (Chapter 6) and the systems approach (Chapter 2) be used to monitor and evaluate implemented innovation policy dealing with these global challenges. Going back a decade, there was an ongoing discussion of the need to initiate innovation policy to deal with the financial crisis of 2008 and to monitor and evaluate policies, once they were implemented, in order to benefit from policy learning which would improve the policy intervention and result in better outcomes (Gault 2010). The same needs are present when dealing with sustainable development and trying to mitigate climate change. The difference since 2008 is the presence of a general definition of innovation that can be used to measure activities related to the SDGs and to climate change, including the linkages between the actors and their impacts on the outcomes of policies. Promoting innovation policy and the measurement of innovation, once the policy was implemented, would increase knowledge of sustainable development and of climate change.

Work on green growth and eco-innovation makes use of innovation, as demonstrated by the OECD Green Growth Strategy (OECD 2011) and the *Maastricht Manual on Measuring Eco-innovation, Green Growth and the Green Economy* (Kemp et al. 2019). Green growth and eco-innovation can happen in any economic sector and links between actors within sectors and across sector boundaries provide information on how policy is functioning. As noted in the text, as part of advancing the understanding of green growth and eco-innovation, the *Maastricht*

Manual needs to find an institutional home where participating countries can use it to guide data gathering and analysis and then to support future developments of the manual.

Innovation and inclusive growth have been part of the innovation measurement discourse for at least a decade and it is covered in this book by 'restricted' innovation introduced in Chapters 5 and 6. While the measure tools are present, work at the OECD (2015d) shows that measurement and evaluation are still developing in the same way as the measurement of innovation in the business sector in the years leading up to the first *Oslo Manual* in 1992. The suggestion is, as with the *Maastricht Manual*, that an institutional home be found to maintain the discourse on inclusive innovation and growth and to support possible manuals in the future

Health issues, including those related to social care, invite innovation policy in health in all sectors and statistical measurement of the results of implemented innovation policy.

There is much to do, but the knowledge and capacity are present in statistical offices and research institutes to do it.

NOTES

1. See 'What is SD' in https://www.un.org/sustainabledevelopment/development-agenda/ (accessed 17 March 2020) or World Commission on Environment and Development (1987).
2. These categories are taken from OECD (2019c). They are not unique and are used to develop the measurement requirements. Other categories could be used (Gault 2018b).
3. See http://www.data4sdgs.org/index.php/about-gpsdd (accessed 17 March 2020).
4. See https://unfccc.int/resource/bigpicture/#content-the-paris-agreemen (accessed 17 March 2020).
5. See https://www.unece.org/fileadmin/DAM/stats/documents/ece/ces/ge.33/2018/mtg4/S2_3_UNSD_CC_rev.pdf (accessed 17 March 2020).
6. See https://ec.europa.eu/environment/green-growth/index_en.htm (accessed 17 March 2020).
7. As an example, see the call for a green growth project in Denmark, https://innovationsfonden.dk/en/programmes/grand-solutions/green-growth.
8. See http://www.centerforhealthsecurity.org/our-work/events/2001_dark -winter/about.html (accessed 17 March 2020) for an experiment which examined the response to the release of a smallpox or similar virus.
9. The digital economy and cybercrime are considered in Chapter 10.

10. Innovation, measurement and policy

10.1 INTRODUCTION

For almost 30 years, official statistics on innovation have come from business surveys. Innovation was seen as a business phenomenon, and the source of jobs and growth. Now consider innovation in all sectors and the interaction of units in all sectors as they engage in innovation. As an example, the results can demonstrate the influence of government innovation on business innovation.

In Chapter 6, the general definition of innovation was introduced, applicable in all economic sectors of the System of National Accounts. The sectors were defined in Chapter 7 where sector-specific innovation was discussed. Chapter 8 dealt with innovation that can appear in any sector, such as innovation in the informal economy. This chapter examines statistical measurement of innovation in any sector and then goes on to discuss innovation policy in this context. The key point is that a general definition of innovation, while fundamental, is just the beginning in understanding measurement and policy in any or all economic sectors.

To start with, the innovation of the institutional unit in each sector can be identified by a survey.[1] The surveys can be business surveys for the business sector or business-like surveys in the case of the non-profit organisation serving households (NPISH) sector and the general government sector (or the public sector). Social surveys, dealing with people, are required to measure innovation in the household sector. This is discussed in Section 10.2.

Another measurement issue is the digital economy and how it affects the measurement of innovation and the collection of data. The rapid development of the digital economy, and its implication for statistical measurement is a factor discussed in Section 10.3.

Section 10.4 considers how to use a systems approach to bring together the measurement of innovation in all economic sectors, starting with agreement on how to use the general definition of innovation and how to measure innovation in each sector.

The remaining sections, 10.5–10.8, deal with innovation policy and Section 10.9 concludes.

10.2   INNOVATION AND DATA COLLECTION

For innovation in the business sector, the use of surveys is well covered by the *Oslo Manual* (OECD/Eurostat 2018) which stresses the importance of a current business register. Business registers are well established in developed countries, but this is not always the case in developing countries. In any country, the business register, maintained by the national statistical office (NSO), may not be available to researchers from outside the statistical office, resulting in the use of commercial registers which might not be as comprehensive or as up to date as the NSO register.

Business registers, and how they capture new firms and detect and remove firms that are no longer trading, is an ongoing activity. The international forum dealing with this subject is the Wiesbaden Group on Business Registers,[2] which is an informal group that reports to the UN Statistical Commission.

When drawing a sample from a business register, a decision is made about the degree of detail that is required for the purpose of the survey. Every entry in the business register should have an industrial classification. An example of such a classification is provided by the International Standard Industrial Classification (ISIC) (UN Statistics Division 2008).[3] ISIC, Revision 4 is divided into sections (one letter of the alphabet), divisions (two digits), groups (three digits) and classes (four digits). If the survey is to report on 'groups', the three-digit level, the sample has to be drawn at the group level with sufficient units to provide population estimates for the groups being studied.

One of the challenges of the digital economy is that units change their activities rapidly while the current version of ISIC, Revision 4 was approved 14 years ago in 2006. The question is whether ISIC, Revision 4 can provide a meaningful classification for studying innovation in all sectors of the economy in 2020. Of course, this question applies to all industrial classifications, not just ISIC.

Table 10.1   *Business characteristics and innovation*

	High entry rate	Low entry rate
High exit rate	*Volatile*	*Structural change*
	New industries	UK retail sector and pubs
Low exit rate	*New opportunities*	*Stable*
	Disruptive innovation	Audit and actuarial services, …

Source: The author.

Another issue is where in the firm the survey should be placed. Should it be at the firm level or the establishment level (or kind-of-activity unit level (KAU))?

For a business register that manages entry and exit well, Table 10.1 could be used to place all N units in the register, classified at the desired level of ISIC, into one of the quadrants. This requires a decision on what is a high level of entry in year Y, and of exit in year Y-1, using the following equation. Once a decision is made on high and low exit and entry, the units could be classified using ISIC codes and then the presence of innovation probed using surveys, or other means of acquiring data.

$$N_Y / N_{Y-1} = 1 + (N_{EnteredY} - N_{ExitedY-1})/ N_{Y-1}$$

Business surveys, or business-like surveys, could then be used to find innovation in the business sector, the NPISH sector and in the public sector (general government sector combined with the aggregate of public corporations (EC et al. 2009: para. 22.41)). For the household sector, social surveys are required and there is also a question about how to draw a sample that represents the population being studied.

Registers of institutional units and of the population are needed for surveys, but there is growing interest in ways of gathering data that do not require a survey. The National Research Council (2014) provides a discussion, and recommendations, on how innovation is measured and how it could be measured, as does the National Academies of Sciences, Engineering and Medicine (2017). These reports come from a panel and a workshop commissioned by the National Center for Science and Engineering Statistics (NCSES) which is part of the National Science Foundation (NSF) in the US and is a statistical office. The reports provide insight into alternative ways of gathering data and ways of providing information on innovation. The key observation is that data collection is expected to change over the next decade and the digital economy is a driving force in this change.

10.3   DIGITAL ECONOMY

The digital economy is a relatively new phenomenon and it has implications for both innovation measurement and innovation policy. At the OECD, the project 'Going Digital: Making the Transformation Work for Growth and Wellbeing' has produced many studies of the digital economy.[4] From the measurement perspective, the key publication is OECD (2019a), which deals with measuring the digital transformation. It makes a call to action because of the speed with which the subject is developing and the need to act now to produce relevant indicators and measurement tools.

Four actions are proposed to produce data that can be used for indicators (OECD 2019a: 17). The actions are to:

1. make the digital transformation visible in economic statistics
2. understand the economic impacts of digital transformation
3. measure wellbeing in the digital age, and
4. design new approaches to data collection.

Five additional actions focus on specific areas. The actions are to:

1. monitor transformative technologies (notably the Internet of Things, AI and Blockchain)
2. make sense of data and data flows
3. define and measure skills needed in the digital era
4. measure trust in online environments, and
5. assess governments' digital strengths.

Action 1, the making of digital transformation visible in economic statistics, echoes the earlier discussion about the need to have innovation, measured in all economic sectors, present in official statistics, which include economic statistics. While Chapter 5 of OECD (2019a) deals with 'Unleashing Innovation', it is more focused on preparing the way for innovation than on the activity of innovation. The first task is to understand the digital economy, and once that has been accomplished, in whatever sector, the question can be raised about measuring innovation.

The International Monetary Fund (IMF 2018) also addresses measuring the digital economy and is reluctant to change the conceptual framework of GDP to include 'free digital services', but it does recommend that 'indicators of welfare' from free digital products can, and should, be developed in the context of non-market production outside of the boundary of the GDP. The IMF goes on to suggest that 'recommendations for overcoming the measurement challenges posed by digitalisation include improving access by national statistics compilers specifically to administrative data and generally "Big Data"'. Free digital services were discussed in Chapter 7. Free product (good or service) innovations have policy implications related to how product producers and users interact.

Implicit in both the OECD and the IMF recommendations is that they are directed at a market economy. This opens a potentially important discussion about innovation in all economic sectors and how it relates, or does not relate, to a market economy. This is where international organisations, such as the International Statistical Institute (ISI), the UN Statistics Division (UNSD), the UNESCO Institute Statistics (UIS), the IMF, the OECD, the World Bank and the World Intellectual Property Organization (WIPO), as well as the EU, a supranational organisation, could help set standards to guide digital economy statistical activities from which inferences could be drawn on innovation activities in the digital economy. Given the rate at which the digital economy is growing, the need for such a consensus is urgent if the innovation in the economy is to be understood.

Paunov and Planes-Satorra (2019: 32) explore how digital technologies are changing innovation in three industries in the business sector, agriculture, the automotive industry and retail. In the course of this study they raise a point about the role of R&D in innovation policy which also has measurement implications. The role of R&D was discussed in Chapter 3 of this book where the point was made that more firms innovate than do R&D and that this is dependent on the (employment) size of the firm. The conclusion is that innovation statistics must go beyond R&D.

In the case of the digital economy, Paunov and Planes-Satorra (2019) note that the following innovation activities might not be captured in R&D investment statistics:

* data and software development innovation activities
* innovation in business processes
* business model innovation, and
* the role of capabilities for innovation.

They observe that, as surveys are conducted at the firm level, they do not capture innovation activities lower down in the structure of the firm, at innovation project level, for example.

In their conclusion, Paunov and Planes-Satorra (2019: 39) note that there are five trends influencing the digital economy. They are that:

- innovation is increasingly data based, enabled also by the deployment of the Internet of Things
- services are at the centre of innovation
- innovation cycles are accelerating
- innovation processes are more collaborative and
- firms invest in new organisational capabilities to better embrace digital innovation.

These trends raise some important questions. The use of data and technologies, where data is functioning as technology, suggest that official statistics should include data on use and planned use of technologies in the digital economy. Recall the discussion of this in Chapter 8, Section 8.4, and the role of user innovation. Services have dominated economies in most countries in the world for at least half a century and it is not a surprise that they play a key role in the digital economy in a world that is dematerialising. The acceleration of innovation cycles make time a key variable in understanding the digital economy and related innovation. Innovation processes are more collaborative in the digital economy, which reflects the complexity of the innovation activities and the need to involve different skills to deliver the innovation. Finally, investing in organisational capabilities is easier than reorganising a manufacturing plant.

With the digital economy, all sectors will be using, or trying to use, digital technologies and they will communicate with one another. To understand innovation in the digital economy, a systems approach is essential.

10.4   SYSTEMS APPROACH

The systems approach to measuring innovation was introduced in Chapter 2 as a means of classifying institutional units in all sectors and their inter-actions, and the role of framework conditions in promoting or impeding innovation in the institutional units. The general definition of innovation in Chapter 6 made it possible to use a standard measure of innovation and support comparisons with innovation activity in all sectors.

The next challenge is to build on previous measurement work to develop guidelines for the measurement of innovation in all economic sectors, using the general definition and the knowledge of measurement developed by researchers on public sector innovation and household sector innovation. Both domains have a long history of measurement but there is no equivalent to the *Oslo Manual* for the general government (public) sector, the NPISH sector or the household sector. Developing these manuals is the first challenge which will take years. The next challenge is to agree on ways of studying the entire innovation system in order to support the development of more effective innovation policy.

10.5   INNOVATION AND POLICY

Innovation policy leading up to the 2020s was discussed in Part II. Chapter 3 reflected the thinking in the recent past about how to promote innovation in the business sector and how the National Innovation Systems approach, and holistic innovation policy, were used to support innovation and to understand it. Chapter 4 provided an overview of how innovation policy, once implemented, could be monitored to demonstrate that it was achieving the desired objectives, or not. Chapter 5 then reviewed how innovation policy was developed in countries and internationally. Also introduced was the concept of 'restricted' innovation which was a means to observe broader policy objectives than just the propensity of an institutional unit (a firm in Part II) to innovate. Restrictions revealed innovation that supported green growth, inclusion, sustainability or other domains of policy interest.

Part III moved the domain of innovation from the business sector to all economic sectors. This was done by applying the general definition of innovation provided in the fourth edition of the *Oslo Manual* (OECD/ Eurostat 2018) and then using the economic sectors in the System of National Accounts (EC et al. 2009) as the domains for sector-specific measurement and analysis, going beyond innovation only in the business sector. However not all innovation is sector specific. Innovation in the informal economy is one example of a type of innovation happening in any or all economic sectors. It requires a different approach.

Moving from innovation in the business sector to innovation in any economic sector is a big step for both measurement and policy development. In parallel with broadening the domain of innovation policy, the digital economy is growing and is changing the way things are done and how innovation happens. In addition, with innovation happening everywhere, and differently, the links between actors in the same sector and with actors in other sectors are more important than they have been. The links can be feedback loops in an innovation system, and they can also be value chains, local or global, including skilled people, data, finance, knowledge, material or technologies. A key consideration is that the measurement and the analysis of linkages of all kinds could result in a deeper understanding of innovation than is available from a business sector perspective, leading to more effective policy interventions.

This discussion of innovation policy in the 2020s, and beyond, starts with the digital economy, then looks at the role of networks in policy development where the networks are made up of the linkages in the innovation system, connecting actors to one another and to framework conditions. This leads to a discussion of how innovation policy and its implementation are to be studied. This is an important issue now that innovation policy has the potential to be more insightful as well as being more complex than when it was directed at a single economic sector.

10.6   DIGITAL ECONOMY AND INNOVATION POLICY

As can be inferred from Paunov and Planes-Satorra (2019), when dealing with digital products and processes, innovation can be more sector specific than dealing with non-digital products and processes (a 'sector' here is a subset of the SNA business sector, such as manufacturing). In Chapter 6, an additional characteristic was discussed as a condition of innovation, the link between the producer and the user that provides the producer with a source of information about what the user is doing with the product. While this linkage is well established, there are unresolved questions about ethics and how the information is used.

The digital products that are provided at prices that are not economically significant, including zero, are raising questions about measurement and how the results fit into official statistics, such as the GDP. In Section 10.3, four actions were presented from OECD (2019a:17) as part of the measurement discussion. These actions also have implications for innovation policy in the digital economy as do the five specific actions in Section 10.3.

10.7   NETWORKS AND INNOVATION POLICY

Chapter 2 described an innovation system that started with actors, the linkages between actors and the outcomes. The system was bounded by framework conditions which could prevent (trade barriers) or encourage (broadband access in all parts of the country) innovation. Measuring innovation in the business sector was a challenge and the measurement of innovation in all economic sectors will be more of a challenge. An important focus will be on linkages of which value chains are examples.

In the digital economy, more attention has to be given to digital platforms and the trust needed to support exchanges in the platform. Such platforms can be created and managed by institutional units in any economic sector and the exchanges made possible on the platforms can be between units in any economic sector. The platforms change with demand and participants. Industry 4.0 is an example that has been reviewed by Horst and Santiago (2018). They also comment upon platforms in developing countries.

10.8   UNDERSTANDING INNOVATION POLICY AND ACTING ON IT

In Chapters 3 and 5, there was discussion of the difficulties of understanding innovation policy and the learning which should result from monitoring and evaluation of implemented policy. These difficulties are not going to be simpler when the policy is expected to deal with more than one economic sector.

The chapters in PART III have demonstrated that innovation is global, complex, dynamic and non-linear in response to policy interventions. Using the systems approach discussed in Chapter 2, policy intervention may or may not result in the intended change as it can be influenced by other policy interventions and connections of the actors in the system with other actors. These connections may include local and global value chains.[5]

Connection to other policy interventions is noted in the EU innovation policy fact sheet of the European Parliament.[6] The description of innovation policy in the fact sheet is that:

> It is also strongly linked to other EU policies, such as those on employment, competitiveness, environment, industry and energy. The role of innovation is to turn research results into new and better services and products in order to

remain competitive in the global marketplace and improve the quality of life of Europe's citizens.

This makes the point that innovation does not happen in isolation and it can be influenced by other policy initiatives, deliberately or not.[7]

Part of the Marburger objective, discussed in Chapter 3, remains, and it is the opinion that the policy process 'would be easier if we had "big models" of the sort economists use to intimidate their adversaries' (Marburger 2007: 30). Further discussion at the OECD Blue Sky Forum in 2006 (Gault 2013: 453) suggested that the minister of research/education/technology/innovation should receive advice comparable to that received by the equivalent of the minister of finance/economy, based on complex and intimidating models such as those used by economists. Finance ministries do indeed use big models of the economy to guide their interventions. At a time when products are being based on data, and there is powerful computing capacity available, there is an opportunity to model the effects of innovation in all economic sectors and how innovation in one sector can influence economic and social activities in other sectors.

Alessandra Colecchia (2007: 297), in summing up the 2006 OECD Blue Sky Forum II, shifted the emphasis of the Marburger proposal from the Ministry of Science and Technology, or of Industry, to the Ministry of Finance, as that was where the money was. Her goal was 'to ensure that the Ministers of Finance, and of the Economy, recognise STI policies as central to the promotion of economic growth and sustainable development'. We are not quite there yet, but the goal remains, as does the need for 'complex and intimidating models'.

10.9   CONCLUSION

The theme throughout this chapter is urgency, economic activity is changing and changing rapidly. Standard industrial classifications may need revision, the digital economy is changing the way innovation happens, and understanding innovation requires data gathering in all economic sectors of the SNA, noting that the means of gathering data is also changing. The suggestion is that internationally agreed manuals be produced to guide the measurement of innovation in all sectors but the business sector, which already has the *Oslo Manual* (OECD/Eurostat 2018), now in its fourth edition. This provides an opportunity for different organisations with subject matter knowledge, and different parts of organisations, to collaborate on providing guidelines for the measurement of innovation in the sector for which they are responsible.

Possible organisations are the ISI and its sub-organisations, the ISO, the International Telecommunication Union (ITU) and the WIPO. Other organisations could be considered.

NOTES

1. There are other means of collecting data discussed in the two US papers, National Research Council (2014) and National Academies of Sciences, Engineering and Medicine (2017).
2. https://unstats.un.org/wiesbadengroup/ (accessed 17 March 2020).
3. Countries may have their own standard industrial classification, and there are classifications covering more than one county. North America, Canada, Mexico and the US have the North American Industry Classification System (NAICS) (Executive Office of the President 2017). The EU has the Nomenclature des Activités Économiques dans la Communauté Européenne (NACE), now in its second revision (Eurostat 2008).
4. To see the OECD Digital Economy Papers and Country Studies, go to http://www.oecd-ilibrary.org (accessed 17 March 2020) and search on 'digital'.
5. The role of value chains in systems of innovation has taken some time to enter the innovation systems discussion. The reader is encouraged to look at 'Global value chains meeting innovation systems: Are there learning opportunities for developing countries' (Pietrobelli and Rabellotti 2011).
6. See http://www.europarl.europa.eu/factsheets/en/sheet/67/innovation-policy (accessed 17 March 2020).
7. The author has observed a country which stopped a major research facility for sound reasons, viewed from one policy perspective, but which was at variance with the policy objective of another government department which was to raise the level of R&D performance in the country. Policy coherence remains a challenge.

11. Conclusion

11.1 INTRODUCTION

This book has reviewed the statistical measurement of innovation in the business sector and the use of the resulting information for the monitoring of implemented innovation policy. As the propensity of firms to innovate in the business sector is not information of high priority to policy makers, 'restricted innovation' was introduced. The restriction could include green or sustainable innovation or inclusive innovation or any other restricted innovation of policy relevance in the business sector. This raised two questions about restricted innovation, the definition of the restriction for measurement purposes and the need to measure the restricted innovation and its outcomes at different points in time if change was being monitored. All of this could be done with the third edition of the *Oslo Manual* (OECD/Eurostat 2005) providing a definition of innovation and of innovation activities for measurement purposes.

Three things have happened since the third edition of the *Oslo Manual* in 2005 was published to guide the statistical measurement of innovation (OECD/Eurostat 2005). The first is that the fourth edition of the *Oslo Manual* was released in 2018 (OECD/Eurostat 2018) with a general definition of innovation, applicable in all economic sectors of the System of National Accounts (SNA) (EC et al. 2009). In the previous three editions of the *Oslo Manual* the presence of innovation 'everywhere' was acknowledged, but then put to one side so that the manual could focus on innovation in the business sector. The difference in 2018 was that the general definition of innovation was actually introduced, but then put to one side to get on with innovation in the business sector.[1]

The second change since 2005 was the use of the SNA economic sectors in the *Oslo Manual*. Prior to 2018 there was the need for only one SNA sector, the business sector, but now, there is a definition of innovation applicable in all economic sectors, such as the general government sector (public sector is the general government sector combined with the aggregate of public corporations) (EC et al. 2009: para. 22.41), the household sector and the non-profit organisations servicing households (NPISH) sector. Over the years there were important experiments in measuring innovation in the public sector and in the household sector, but without a common definition. It is now possible to bring the measurement of innovation in all sectors together to make comparisons and to see differences, while knowing that the underlying definition of innovation was consistent in all sectors.

The third change was the rapid development of the digital economy. As products and processes became digital, along with their delivery and maintenance, how innovation happened started to change. This is becoming a radical transformation in the economy and society, present in all economic sectors.

With the general definition and in the economic sectors of the SNA, innovation can be measured everywhere but with the growth of the digital economy, products and processes are changing as is the way innovation happens. All of this change affects innovation policy and how it is implemented. As we enter the 2020 decade, there is work to be done.

11.2   AN AGENDA

Given that innovation measurement and policy are changing rapidly, and that innovation contributes significantly to the economy and the society, some questions are posed in what follows for the consideration of the reader. The allocation of resources to innovation measurement and policy is a measure of their importance.

11.2.1   Official Statistics

In 2020, there are statistics produced by governments through national statistical offices or by contracting with research institutions. Since the first edition of the *Oslo Manual* (OECD 1992) the statistical measurement was governed by the *Oslo Manual* and the results published as official statistics from the business sector for use by governments and publication in scoreboards to support international comparison. One of the uses could be the adjustment of existing innovation policy to make it more effective, hence the importance of monitoring and evaluation.

With a general definition of innovation this process can be extended to any economic sector. What is missing is the equivalent of the *Oslo Manual* for the other sectors. What makes the *Oslo Manual* successful is that it is a product of the OECD and Eurostat and it is developed by the delegates, including the European Union/Eurostat, of the OECD Working Party of National Experts on Science and Technology Indicators (NESTI). NESTI is not limited to indicators in the business sector, as shown by the *Frascati Manual* (OECD 2015c), which covers the performance of R&D in all economic sectors. It may be time to resume work on a possible manual for public sector innovation.

The question remains as to whether countries want an internationally standard manual that provides guidance for the measurement of innovation in the public sector. This question extends to manuals for innovation in the household and the NPISH sectors. Related questions are how innovation policy is used in these sectors and whether the statistical measurement of innovation influences policy development.

11.2.2   Beyond Sectors

While there is sectoral innovation, some kinds of innovation can happen in any sector, an example being innovation in the informal economy. Both statistical measurement and policy implementation are affected by the lack of information on institutional units engaged in the informal economy. This raises methodological issues: how to find the units and whether they are engaged in innovation.

In African countries, informal activities can account for a significant portion of GDP and innovation in the informal economy; both its detection through measurement and its influence through policy (Chapter 8) are important to the economy and society.

Other examples of innovation that can appear in any sector are eco-innovation (the *Maastricht Manual* is part of this discussion) and innovation resulting from the use of general purpose technologies. There is also social innovation.

Should innovation measurement and policy in these areas be seen as experimental, as the measurement of business sector innovation was for over a decade before the first edition of the *Oslo Manual* changed the situation? Or should there be international standards and guidelines?

11.2.3   Digital Economy

The digital economy is growing (more digital products) and product and process innovation differs from material product and process innovation. As discussed in Chapter 7, digital products can be made available at prices that are not economically significant, including zero price products. In the discussion of zero price products a third condition for innovation was introduced. The first was 'a new or improved product that differs significantly from the unit's previous products', the second was 'that has been made available to potential users', and the suggested third was to 'have a connection between the user and the producer'. The third condition could be added as a restriction on the measurement of the product innovation.

The question is whether more effort should go into understanding innovation measurement and policy in the digital economy (OECD 2019a), which is present in all economic sectors.

11.3   THE END

The purpose of this book is to engage the reader in the role of innovation measurement and the use of the resulting information in the development of policy and in the monitoring and evaluation of implemented innovation policy. A recurring message is that innovation measurement and innovation policy are not simple but both are important to understanding the economy and society in which we live.

NOTE

1.   The general definition in the fourth edition of the *Oslo Manual* (OECD/ Eurostat 2018) is found in Chapter 1, para. 1.25, and in chapter 2, para. 2.99. The definition of innovation in the business sector is found in Chapter 3, para. 3.9. From there on in Chapters 3–11 of the *Oslo Manual*, the focus is on business sector innovation. This is consistent with the history of the *Oslo Manual* which was always focused on the business sector.

References

Aho, Esko, Mikko Alkio and Ilkka Lakaniemi (2013), The Finnish approach to innovation strategy and indicators, in Fred Gault (ed.), *Handbook of Innovation Indicators and Measurement*, Cheltenham, UK and Northampton, MA, USA: Edward Elgar, pp. 320–32.

Appelt, Silvia, Fernando Galindo-Rueda and Ana Cinta González Cabral (2019), Measuring R&D tax incentives: The new OECD R&D Tax Incentive Database, *OECD Science, Technology and Industry Working Papers*, Paris: OECD Publishing.

Arundel, Anthony and D. Huber (2013), From too little to too much innovation? Issues in measuring innovation in the public sector, *Structural Change and Economic Dynamics*, **27**, 146–59.

Arundel, Anthony and Keith Smith (2013), History of the Community Innovation Survey, in Fred Gault (ed.), *Handbook of Innovation Indicators and Measurement*, Cheltenham, UK and Northampton, MA, USA: Edward Elgar, pp. 60–87.

Arundel, Anthony, Luca Casali and Hugo Hollanders (2015), How European public sector agencies innovate: The use of bottom-up, policy-dependant and knowledge-scanning innovation methods, *Research Policy*, **44**, 1271–82.

Arundel, Anthony, Carter Bloch and Barry Ferguson (2019), Advancing innovation in the public sector: Aligning innovation measurement with policy goals, *Research Policy*, **48**, 789–98.

AU (2014), *Science, Technology and Innovation Strategy for Africa, STISA 2024*, Addis Ababa: African Union.

AU-NEPAD (2010), *African Innovation Outlook 2010*, Pretoria: AU-NEPAD.

AUDA-NEPAD (2019), *African Innovation Outlook 2019*, Pretoria: AUDA-NEPAD.

Bloch, Carter (2010a), *Measuring Pubic Innovation in the Nordic Countries: Final Report*, Aarhus: The Danish Centre for Studies in Research and Research Policy.

Bloch, Carter (2010b), *Towards a Conceptual Framework for Measuring Public Sector Innovation, Module 1 – Conceptual Framework*, Aarhus: The Danish Centre for Studies in Research and Research Policy.

Bloch, Carter (2013), Measuring innovation in the public sector, in Fred Gault (ed.), *Handbook of Innovation Indicators and Measurement*, Cheltenham, UK and Northampton, MA, USA: Edward Elgar, pp. 403–19.

Bloch, C. and M. Bugge (2013), Public sector innovation – from theory to measurement, *Structural Change and Economic Dynamics*, **27**, 133–45.

BMWi (2019), *Financing Start-ups and Growth: Overview of Funding Instruments*, Berlin: BMWi.

Borrás, Susana and Charles Edquist (2019), *Holistic Innovation Policy: Theoretical Foundations, Policy Problems and Instrument Choices*, Oxford: Oxford University Press.

Brynjolfsson, Erik, Avinash Collis, Erwin Diewert, Felix Eggers and Kevin Fox (2019), GDP-B: Accounting for the value of new and free goods in the digital economy, Discussion Paper, Vancouver School of Economics, Vancouver: University of British Columbia.

Chaminade, Cristina, Bengt-Åke Lundvall and Shagufta Haneef (2018), *Elgar Advanced Introductions, Advanced Introduction to National Innovation Systems*, Cheltenham, UK and Northampton, MA, USA: Edward Elgar.

Charmes, Jacques (2016), The informal economy: definitions, size, contribution and main characteristics, in Erika Kraemer-Mbula and Sacha Wunsch-Vincent (eds), *The Informal Economy in Developing Nations – Hidden Engine of Innovation*, Cambridge: Cambridge University Press, pp. 13–44.

Charmes, Jacques (2019), *Dimensions of Resilience in Developing Countries: Informality, Solidarities and Care Work*, Cham, Switzerland: Springer Nature Switzerland AQ.

Charmes, Jacques, Fred Gault and Sacha Wunsch-Vincent (2016), Formulating an agenda for the measurement of innovation in the informal economy, in Erika Kraemer-Mbula and Sacha Wunsch-Vincent (eds), *The Informal Economy in Developing Nations – Hidden Engine of Innovation*, Cambridge: Cambridge University Press, pp. 332–62.

Charmes, Jacques, Fred Gault and Sacha Wunsch-Vincent (2018), Measuring innovation in the informal economy – formulating an agenda for Africa, *Journal of Intellectual Capital*, **19**(3), 536–49. https://doi.org/10.1108/JIC-11-2016 -0126

Colecchia, A. (2007), Looking ahead: What implications for STI indicator development, in OECD (ed.), *Science, Technology and Innovation Indicators in a Changing World: Responding to Policy Needs*, Paris: OECD Publishing, pp. 285–98.

Cornell University, INSEAD and WIPO (2019), *The Global Innovation Index 2019: Creating Healthy Lives – the Future of Innovation*, Ithaca, NY, Fontainebleau and Geneva.

Coyle, Diane (2016), *GDP, a Brief but Affectionate History*, Princeton, NJ: Princeton University Press.

Cozzens, S. and D. Thakur (eds) (2014), *Innovation and Inequality: Emerging Technologies in an Unequal World*, Cheltenham, UK and Northampton, MA, USA: Edward Elgar.

Davis, K.E., A. Fischer, B. Kingsbury and S. Merry (eds) (2012), *Governance by Indicators: Global Power through Quantification and Rankings*, Oxford: Oxford University Press.

de Jong, Jeroen P.J., Eric von Hippel, Fred Gault, Jari Kuusisto and Christina Raasch (2015), Market failure in the diffusion of consumer-developed innovations: Patterns in Finland, *Research Policy*, **44**(10), 1856–65. http://www .sciencedirect.com/science/article/pii/S0048733315001122 (accessed 17 March 2020).

Diewert, Erwin, Kevin Fox and Paul Schreyer (2018), Experimental econom-
ics and the new goods problem, Discussion Paper, Vancouver School of
Economics, Vancouver: University of British Columbia.
Drummond, Don and Alistair Bentley (2010), The Productivity Puzzle: Why is
the Canadian record so poor and what can be done about it? Toronto: TD Bank
Financial Group. https://www.td.com/document/PDF/economics/special/td
-economics-special-ab0610-productivity.pdf (accessed 17 March 2020).
Dutrénit, Gabriela and Judith Sutz (2014), *National Innovation Systems, Social
Inclusion and Development, the Latin American Experience*, Cheltenham, UK
and Northampton, MA, USA: Edward Elgar.
EC (2014), *European Public Sector Innovation Scoreboard 2013*, Brussels:
Publications Office of the EU.
EC, IMF, OECD, UN and the World Bank (1994), *System of National Accounts,
1994 (1994 SNA)*, New York: United Nations.
EC, IMF, OECD, UN and the World Bank (2009), *System of National Accounts,
2008 (2008 SNA)*, New York: United Nations.
Edler, Jakob and Jan Fagerberg (2017), Innovation policy: What, why, and how?
Oxford Review of Economic Policy, **33**(1), 2–23.
Edler, Jacob, Paul Cunningham, Abdullah Gök and Philip Shapira (2016),
Handbook of Innovation Policy Impact, Cheltenham, UK and Northampton,
MA, USA: Edward Elgar.
Edquist, C. (ed.) (1997), *Systems of Innovation: Technologies, Institutions and
Organisations*, London: Pinter.
Edquist, C. (2005), Systems of innovation: Perspectives and challenges, in Jan
Fagerberg, David Mowery and Richard Nelson (eds), *The Oxford Handbook
of Innovation*, Oxford: Oxford University Press, pp. 181–208.
Edquist, C. and L. Hommen (eds) (2008), *Small Country Innovation Systems,
Globalisation, Change and Policy in Asia and Europe*, Cheltenham, UK and
Northampton, MA, USA: Edward Elgar.
Edquist C. and J.M. Zabala-Iturriagagoitia (2018), Viewpoint: The latest EU
innovation index is out. It's flawed, *Science Business*, 22 June, London:
Science Business. https://sciencebusiness.net/viewpoint/viewpoint-latest-eu
-innovation-index-out-its-flawed (accessed 17 March 2020).
EFI-Commission of Experts for Research and Innovation (2017), *Report on
Research, Innovation and Technological Performance in Germany 2017*,
Berlin: EFI.
EFI-Commission of Experts for Research and Innovation (2019), *Report on
Research, Innovation and Technological Performance in Germany 2019*,
Berlin: EFI.
Ergas, H. (1986), Does technology policy matter? *CEPS Papers* No. 29, Brussels:
Centre for European Studies.
Executive Office of the President (2017), *North American Industry Classification
System (NAICS)*, Washington, DC: Executive Office of the President.
EU (2013), *Lessons from a Decade of Innovation Policy*, Brussels: European
Union.
EU (2014), *Boosting Open Innovation and Knowledge Transfer in the European
Union*, Brussels: Publication Office of the European Union.

European Commission (2019), *European Innovation Scoreboard 2019*, Brussels: Publication Office of the European Union.

Eurostat (2008), *NACE.Rev.2, Statistical Classification of Economic Activities in the European Community*, Luxembourg: Office for Official Publications of the European Communities.

Eurostat (2018), *Europe 2020 Indicators – Poverty and Social Exclusion, Statistics Explained*, Brussels: Publication Office of the European Union. https://ec.europa.eu/eurostat/statistics-explained/index.php/Europe_2020 _indicators_-_poverty_and_social_exclusion (accessed 17 March 2020).

Fagerberg, Jan, David Mowery and Richard Nelson (2005), *The Oxford Handbook of Innovation*, Oxford: Oxford University Press.

Foray, Dominique and Fred Gault (eds) (2003), *Measuring Knowledge Management in the Business Sector: First Steps*, Paris: OECD Publishing.

Francis-Devine, Brigid, Lorna Booth and Feargal McGuinness (2019), Poverty in the UK: Statistics, Briefing Paper No. 7096, London: House of Commons Library.

Freeman, Chris (1987), *Technology Policy and Economics Performance: Lessons from Japan*, London: Pinter.

Freeman, Chris and Luc Soete (2007), Developing Science, Technology and Innovation Indicators: The twenty-first century challenges, in OECD (ed.), *Science, Technology and Innovation Indicators in a Changing World, Responding to Policy Needs*, Paris: OECD Publishing, pp. 271–80.

Gault, F. (2010), *Innovation Strategies for a Global Economy, Development, Implementation, Measurement and Management*, Cheltenham, UK and Northampton, MA, USA: Edward Elgar and Ottawa: IDRC.

Gault, Fred (2011), Developing a science of innovation policy internationally, in Kaye Husbands-Fealing, Julia Lane, John Marburger and Stephanie Shipp (eds), *Science of Science Policy: A Handbook*, Stanford, CA: Stanford University Press, pp. 156–82.

Gault, Fred (2012), User innovation and the market, *Science and Public Policy*, **39**, 118–28.

Gault, Fred (ed.) (2013), *Handbook of Innovation Indicators and Measurement*, Cheltenham, UK and Northampton, MA, USA: Edward Elgar.

Gault, F. (2014), Where are innovation indicators, and their applications, going? UNU-MERIT Working Paper 2014-055, Maastricht: UNU-MERIT, 19 pp.

Gault, Fred (2015), Measuring innovation in all sectors of the economy, UNU-MERIT Working Paper 2015-038, 23 pp.

Gault, Fred (2016), User innovation and official statistics, in Dietmar Harhoff and Karim R. Lakhani (eds), *Revolutionizing Innovation: Users, Communities and Open Innovation*, Cambridge, MA: The MIT Press, pp. 89–105.

Gault, F. (2018a), Defining and measuring innovation in all sectors of the economy, *Research Policy*, **47**, 617–22.

Gault, F. (2018b), Measuring the economic and social impact of innovation for sustainable development, in Nuria Sanz and Carlos Tejada (eds), *Innovación para el Dessarrollo Sostenible*, México: Gobierno del Estado de Guanajuato en colaboración con la Oficina de la UNESCO en México, pp. 173–80. http://

unesdoc.unesco.org/images/0026/002656/265693m.pdf (accessed 17 March 2020).

Gault, F. (2019), User innovation in the digital economy, *Foresight and STI Governance*, **13**(3), 6–13.

Gault, F. and M. Hakvåg (2018), *Cooperation between the International Organization for Standardisation (ISO) and the Organisation for Economic Co-operation and Development (OECD) of the Definition of Innovation for International Management and Statistical Measurement*, Paris: OECD. https://community.oecd.org/docs/DOC-165568 (accessed 17 March 2020).

George, Gerard, Ted Baker, Paul Tracy and Havovi Joshi (2019), Inclusion and innovation: A call to action, in Gerard George, Ted Baker, Paul Tracy and Havovi Joshi (eds), *Handbook of Inclusive Innovation, the Role of Organisations, Markets and Communities in Social Innovation*, Cheltenham, UK and Northampton, MA, USA: Edward Elgar, pp. 2–22.

Ghafur, S., G. Fontana, G. Martin, A. Grass, J. Goodman and A. Darzi (2019), *Improving Cyber Security in the NHS*, Institute of Global Health Innovation, London: Imperial College.

GTIPA (2019), *National Innovation Policies: What Countries Do Best and How They Can Improve*, Washington DC: Global Trade & Innovation Policy Alliance.

Guellec, Dominique and Caroline Paunov (2018), Innovation policies in the digital economy, *OECD Science, Technology and Innovation Policy Papers*, No. 59, November, Paris: OECD Publishing.

Harhoff, Dietmar and Karime Lakhani (eds) (2016), *Revolutionizing Innovation: Users, Communities and Open Innovation*, Cambridge, MA: The MIT Press.

Haskel, Jonathan and Stian Westlake (2018), *Capitalism without Capital*, Princeton, NJ: Princeton University Press.

Hawkins, Richard, Knut Blind and Robert Page (2017), *Handbook of Innovation Standards*, Cheltenham, UK and Northampton, MA, USA: Edward Elgar.

Hill, Christopher T. (2013), US innovation strategy and policy: An indicators perspective, in Fred Gault (ed.), *Handbook of Innovation Indicators and Measurement*, Cheltenham, UK and Northampton, MA, USA: Edward Elgar, pp. 333–46.

Hollanders, Hugo and Norbert Janz (2013), Scoreboard and indicator reports, in Fred Gault (ed.), *Handbook of Innovation Indicators and Measurement*, Cheltenham, UK and Northampton, MA, USA: Edward Elgar, pp. 279–97.

Horst, J. and F. Santiago (2018), *What Can Policy Makers Learn from Germany's Industrie 4.0 Development Strategy?* Vienna: UNIDO.

House of Commons Library Briefing Paper (2019), NHS Key Statistics: England, February, Briefing Paper 7281, London: House of Commons Library.

Howaldt, J., A. Butzin, D. Domanski and C. Kaletka (2014), *Theoretical Approaches to Social Innovation – a Critical Literature Review*. A deliverable of the project: 'Social Innovation: Driving Force of Social Change' (SI-DRIVE), Dortmund: Sozialforschungsstelle.

IMF (2018), *Measuring the Digital Economy*, Washington, DC: IMF.

Kemp, R., A. Arundel, C. Rammer, M. Miedzinski, C. Tapia, N. Barbieri, S. Türkeli, A.M. Bassi, M. Mazzanti, D. Chapman, F. Diaz López, W.

McDowall (2019), *Maastricht Manual on Measuring Eco-Innovation for a Green Economy*, Maastricht: University of Maastricht.

Kindlon, Audrey E. and John E. Jankowski (2017), *Rates of Innovation among U.S. Businesses Stay Steady: Data from the 2014 Business R&D and Innovation Survey*, National Center for Science and Engineering Statistics, InfoBrief, NSF 17-321, Arlington, VA: NSF.

Kraemer-Mbula, Erika and Watu Wamae (2010), *Innovation and the Development Agenda*, Paris: OECD Publishing.

Kraemer-Mbula, Erika and Sacha Wunsch-Vincent (eds) (2016), *The Informal Economy in Developing Nations – Hidden Engine of Innovation*, Cambridge: Cambridge University Press.

Kuusisto, Jari, Jeroen P.J. de Jong, Fred Gault, Christina Raasch and Eric von Hippel (2013), Consumer innovation in Finland: Incidence, diffusion and policy implications, *Proceedings of the University of Vaasa Reports 189*, Vaasa, Finland: University of Vaasa.

Lipsey, R.G, K.I. Carlaw and C.T. Bekar (2005), *Economic Transformations, General Purpose Technologies and Long Term Economic Growth*, Oxford: Oxford University Press.

Lundvall, B.-Å. (ed.) (1992), *National Innovation Systems: Towards a Theory of Innovation and Interactive Learning*, London: Pinter.

Lundvall, B.-Å., K.J. Joseph, Cristina Chaminade and Jan Vang (eds) (2009), *Handbook of Innovation Systems and Developing Countries, Building Domestic Capabilities in a Global Setting*, Cheltenham, UK and Northampton, MA, USA: Edward Elgar.

Marburger, John (2005), 'Wanted: Better benchmarks', *Science*, **308**(5725), May, 1087.

Marburger, John (2007), The Science of Science and Innovation Policy, in OECD (ed.), *Science, Technology and Innovation Indicators in a Changing World, Responding to Policy Needs*, Paris: OECD Publishing, pp. 27–32.

Marburger, John and Robert R. Crease (eds) (2015), *Science Policy Up Close*, Cambridge, MA: Harvard University Press.

Marée, Michel and Sybille Mertens (2012), The limits of economic value in measuring the performance of social innovation, in Alex Nicholls and Alex Murdock (eds), *Social Innovation: Blurring Boundaries to Reconfigure Markets*, Basingstoke, UK: Palgrave Macmillan, pp. 114–36.

Mashelkar, R.A. (2012), *On Building an Inclusive Innovation Ecosystem*, Paris: OECD Publishing. http://www.oecd.org/sti/inno/k_mashelkar.pdf (accessed 17 March 2020).

Mashelkar, R.A. (2014), Accelerated inclusive growth through inclusive innovation, presentation at the OECD-Growth Dialogue Symposium on Innovation and Inclusive Growth, Paris, 20 March. http://www.oecd.org/sti/inno/Session _3_Mashelkar_Keynote.pdf (accessed 17 March 2020).

Mazzucato, Mariana (2013), *The Entrepreneurial State, Debunking Public vs. Private sector Myths*, London: Anthem Press.

Meadows, Donella, H. and Diana Wright (eds) (2008), *Thinking in Systems: A Primer*, Sterling, VT: Chelsea Green Publishing.

Ministry of Education, Science and Technology, Kenya (2016), *The Kenya Innovation Indicators Survey 2015*, Nairobi: Ministry of Education, Science and Technology.

Mitchell, Melanie (2009), *Complexity, a Guided Tour*, New York: Oxford University Press.

Molotja, N., S. Parker and P. Mudavanhu (2019), Patterns of investing in business R&D in South Africa, *Foresight and STI Governance*, **13**(3), 51–60.

Moulaert, Frank and Diana MacCallum (2019), *Advanced Introduction to Social Innovation*, Elgar Advanced Introductions series, Cheltenham, UK and Northampton, MA, USA: Edward Elgar.

Moulaert, Frank, Diana MacCallum, Abid Mehmood and Abdelillah Hamdouch (eds) (2013), *The International Handbook on Social Innovation: Collective Action, Social Learning and Transdisciplinary Research*, Cheltenham, UK and Northampton, MA, USA: Edward Elgar.

Mulgan, Geoff (2007), 'What social innovation is', in Geoff Mulgan, Simon Tucker, Rushanara Ali and Ben Sanders (eds), *Social Innovation, What It Is, Why It Matters, and How It Can be Accelerated*, Oxford: Skoll Centre for Social Entrepreneurship, Oxford Said Business School, pp. 8–12.

Mulgan, Geoff (2019), *Social Innovation: How Societies Find the Power to Change*, Bristol: Policy Press.

Mulgan, Geoff, Kippy Joseph and Will Norman (2013), Indicators for social innovation, in Fred Gault (ed.), *Handbook of Innovation Indicators and Measurement*, Cheltenham, UK and Northampton, MA, USA: Edward Elgar, pp. 420–37.

Murray, Robin, Julie Caulier-Grice and Geoff Mulgan (2010), *The Open Book of Social Innovation*, London: The Young Foundation and Nesta.

National Academies of Sciences, Engineering and Medicine (2017), *Advancing Concepts and Models for Measuring Innovation: Proceedings of a Workshop*, Washington, DC: The National Academies Press.

National Research Council (2014), *Capturing Change in Science, Technology and Innovation: Improving Indicators to Inform Policy*, Washington, DC: The National Academies Press.

National Science Board (2018), *Science and Engineering Indicators*, Arlington, VA: NSF.

Nelson, Richard R. (ed.) (1993), *National Systems of Innovation*, New York: Oxford University Press.

Nicholls, A. and A. Murdock (eds) (2012), *Social Innovation: Blurring Boundaries to Reconfigure Markets*, London: Macmillan.

North, D. (1990), *Institutions, Institutional Change and Economic Performance*, Cambridge: Cambridge University Press.

NPCA (2014), *African Innovation Outlook II*, Pretoria: NEPAD Planning and Coordinating Agency.

NSF (2019), Dear Colleague Letter: 2019 Social, Behavioural and Economic (SBE) Repositioning, NSF 19-2019, Arlington, VA: NSF.

OECD (1992), *OECD Proposed Guidelines for Collecting and Interpreting Technological Innovation Data – Oslo Manual*, OCDE/GD (92) 26, Paris: OECD Publishing.

OECD (2002), *Frascati Manual: Proposed Standard Practice for Surveys on Research and Development*, Paris: OECD Publishing.

OECD (2007), *Science, Technology and Innovation Indicators in a Changing World: Responding to Policy Needs*, Paris OECD Publishing.

OECD (2008), *Growing Unequal? – Income Distribution and Poverty in OECD Countries*, Paris: OECD Publishing.

OECD (2009), *Clusters, Innovation and Entrepreneurship*. Paris: OECD Publishing.

OECD (2010a), *The OECD Innovation Strategy: Getting a Head Start on Tomorrow*, Paris: OECD Publishing.

OECD (2010b), *Measuring Innovation: A New Perspective*, Paris: OECD Publishing.

OECD (2011), *Business Innovation Policies: Selected Country Comparisons*, Paris: OECD Publishing.

OECD (2014), *OECD Science, Technology and Industry Outlook 2014*, Paris: OECD Publishing.

OECD (2015a), *The Innovation Imperative: Contributing to Productivity, Growth and Well-being*, Paris: OECD Publishing.

OECD (2015b), *The Innovation Imperative in the Public Sector: Setting an Agenda for Action*, Paris: OECD Publishing.

OECD (2015c), *Frascati Manual 2015, Guidelines for Collecting and Reporting Data on Research and Experimental Development*, Paris: OECD Publishing.

OECD (2015d), *Innovation Policies for Inclusive Growth*, Paris: OECD Publishing.

OECD (2016), *OECD Science, Technology and Innovation Outlook 2016*, Paris: OECD Publishing.

OECD (2017a), *OECD Reviews of Innovation Policy, Kazakhstan*, Paris: OECD Publishing.

OECD (2017b), *OECD Reviews of Innovation Policy, Norway*, Paris: OECD Publishing.

OECD (2017c), *OECD Reviews of Innovation Policy, Finland*, Paris: OECD Publishing.

OECD (2017d), *OECD Science, Technology and Industry Scoreboard 2017*, Paris: OECD Publishing.

OECD (2017e), *OECD Digital Economy Outlook 2017*, Paris: OECD Publishing.

OECD (2018a), *OECD Reviews of Innovation Policy, Austria*, Paris: OECD Publishing.

OECD (2018b), *OECD Science, Technology and Innovation Outlook 2018*, Paris: OECD Publishing.

OECD (2019a), *Measuring the Digital Transformation: A Roadmap for the Future*, Paris: OECD Publishing.

OECD (2019b), Portugal 2019: Review of higher education, research and innovation, *OECD Reviews of Innovation Policy*, Paris: OECD Publishing.

OECD (2019c), *Measuring Distance to the SDG Targets 2019, an Assessment of Where OECD Countries Stand*, Paris: OECD Publishing.

OECD (2019d), *Digital Innovation, Seizing Policy Opportunities*, Paris: OECD Publishing.

OECD/Eurostat (1997*)*, *Proposed Guidelines for Collecting and Interpreting Technological Innovation Data, Oslo Manual*, Paris: OECD Publishing.

OECD/Eurostat (2005), *Oslo Manual, Guidelines for Collecting and Interpreting Innovation Data*, Paris: OECD Publishing.

OECD/Eurostat (2018), *Oslo Manual 2018: Guidelines for Collecting, Reporting and Using Data for Innovation*, 4th edition, The Measurement of Scientific, Technological and Innovation Activities, Paris: OECD Publishing and Luxembourg: Eurostat.

Osborne, Stephen and Louise Brown (2013), *Handbook of Innovation in Public Services*, Cheltenham, UK and Northampton, MA, USA: Edward Elgar.

Paunov, Caroline and Sandra Planes-Satorra (2019), How are digital technologies changing innovation? Evidence from agriculture, the automotive industry and retail, *OECD Science, Technology and Industry Policy Papers*, No. 74, Paris: OECD Publishing.

Pietrobelli, C. and R. Rabellotti (2011), Global value chains meeting innovation systems: Are there learning opportunities for developing countries, *World Development*, **39**(7), 1261–9.

Planes-Satorra, Sandra and Caroline Paunov (2017), Inclusive innovation policies. Lessons from international case studies, *OECD Science, Technology and Industry Policy Papers*, 2017/02, Paris: OECD Publishing.

Rammer, C. and T. Schubert (2016), *Concentration on the Few? R&D and Innovation in German Firms 2001 to 2013*, Mannheim: ZEW, Centre for European Economic Research. http://ftp.zew.de/pub/zew-docs/dp/dp16005.pdf (accessed 17 March 2020).

RICYT/OECD/CYTED (2001), *Standardization of Indicators of Technological Innovation in Latin American and Caribbean Countries: Bogotá Manual*, Buenos Aires: RICYT.

Rüede, Dominik and Kathrin Lurtz (2012), Mapping the various meanings of social innovation: Towards a differentiated understanding of an emerging concept, EBS Business School Research Paper Series 12-03, Oestrich-Winkel: EBS.

Schellings, R. and F. Gault (2002), *Size and persistence of R&D performance in Canadian Firms 1994 to 2002*, Catalogue 88F0006XIE, No. 008, Ottawa: Statistics Canada.

Schumpeter, J. (1934), *The Theory of Economic Development*, Cambridge, MA: Harvard University Press.

Schwab, Klaus (2017), *The Fourth Industrial Revolution*, UK: Portfolio Penguin.

Schwab, Klaus (2018), *Shaping the Future of the Fourth Industrial Revolution, a Guide to Building a Better World*, UK: Portfolio Penguin.

Smith, K. (2005), Measuring innovation, in J. Fagerberg, D.C. Mowery and R.R. Nelson (eds), *The Oxford Handbook of Innovation*, Oxford: Oxford University Press, pp. 148–77.

Smits, Ruud, Stefan Kuhlmann and Philip Shapira (eds) (2010), *The Theory and Practice of Innovation Policy*, Cheltenham, UK and Northampton, MA, USA: Edward Elgar.

Soete, Luc, Bart Verspagen and Bas Ter Weel (2010), Systems of Innovation, in Bronwyn H. Hall and Nathan Rosenberg (eds), *Economics of Innovation*, Amsterdam: North-Holland, Vol. 2, pp. 1159–80.

Teich, Albert H. (2018), In search of evidence-based science policy: From the endless frontier to SciSIP, *Annals of Science and Technology Policy*, **2**(2), 75–199.

UN Statistics Division (2008), *International Standard Industrial Classification of All Economic Activities, Revision 4*, New York: United Nations.

UN Statistics Division (2018), *Global Work on Climate Change Statistics and Indicators and Adaptation-Related SDG Indicators*, Fifth Meeting of the Expert Group on Environment Statistics, New York: United Nations.

UN Statistics Division (2019a), *Global Indicator Framework for the Sustainable Development Goals and Targets of the 2030 Agenda for Sustainable Development*, A/RES/71313, E/CH.3/2018/2, E/CN.3/2019/2, New York: United Nations. https://unstats.un.org/sdgs/indicators/Global%20Indicator %20Framework%20after%202019%20refinement_Eng.pdf (accessed 17 March 2020).

UN Statistics Division (2019b), *Sustainable Development Goal (SDG) Indicators Correspondence with the Basic Set of Environment Statistics of the FDES 2013*, Environment Statistics, 12 July, New York: United Nations. https:// unstats.un.org/unsd/envstats/fdes/SDGsInd_BasicSetMatrix.pdf (accessed 17 March 2020).

United Nations (2015), *Transforming Our World: The 2030 Agenda for Sustainable Development*, A/RES/70/1, New York: United Nations. https:// sustainabledevelopment.un.org (accessed 17 March 2020).

United Nations (2017), *Framework for the Development of Environment Statistics*, New York: United Nations.

Uyarra, E. and R. Ramlogan (2016), The impact of cluster policy on innovation, in J. Edler, P. Cunningham, A. Gök and P. Shapira (eds), *Handbook of Innovation Policy Impact*, Cheltenham, UK and Northampton, MA, USA: Edward Elgar, pp. 196–238.

von Bogdandy, A. and M. Goldmann (2012), Taming and framing indicators: A legal reconstruction of the OECD's programme for international student assessment (PISA), in K.E. Davis, A. Fischer, B. Kinsbury and S. Merry (eds), *Governance by Indicators: Global Power through Quantification and Rankings*, Oxford: Oxford University Press, pp. 52–85.

von Hippel, Eric (1988), *The Sources of Innovation*, New York: Oxford University Press.

von Hippel, Eric (2005), *Democratizing Innovation*, Cambridge, MA: The MIT Press.

von Hippel, Eric (2007), Democratizing innovation: The evolving phenomenon of user innovation, in OECD (ed.), *Science, Technology and Innovation in a Changing World, Responding to Policy Needs*, Paris: OECD, pp. 125–38.

von Hippel, Eric (2016), *Novel Policies Required to Support Free Household Sector Innovation*, OECD Blue Sky Forum III. http://www.oecd.org/sti/blue -sky-2016-agenda.htm#ps4_d2 (accessed 17 March 2020).

von Hippel, Eric (2017), *Free Innovation*, Cambridge, MA: The MIT Press.

World Commission on Environment and Development (1987), *Our Common Future*, Oxford: Oxford University Press.

Index

activities
    in systems 7, 8
    see also innovation activities;
        innovative activities
actors *see* institutional units
Africa 22, 88, 104
African Innovation Outlook (AIO)
    31, 54
African Union 45, 54
ageing population 22
agenda for action 103–5
    beyond sectors 104
    digital economy 104–5
    official statistics 103–4
apps 2–3, 73
artificial intelligence (AI) 3, 8, 21,
    40, 73
Arundel, A. 61, 64
Australia 64
Austria 31

Balamatsias 75
behaviour change 3, 8, 21–2
Bentley, A. 23
'better', in public sector innovation 64
big data 73, 95
big models 100
Blue Sky Forum I (1996) 48
Blue Sky Forum II (2006) 15, 16, 42,
    100
Blue Sky Forum III (2016) xi, 115
Bogotá Manual 47
Borrás, S. 10, 11, 42
'brought into use' 3, 16, 32, 49, 50,
    61, 64, 66, 67, 72, 76, 82
Bulgaria 34
business model(s) 73, 95
Business R&D and Innovation Survey
    (BRDIS) 20

business registers 92
business sector
    defined 61
    R&D performers 23
    surveys 91, 92, 93
business sector innovation
    digital technologies 95
    EU support for 19
    high and low entry rates 93
    and the market 61–2
    measurement 36, 61–2, 87
    Oslo Manual definitions 36, 43,
        49, 50, 51, 61
    policy development 40–45
    see also firm-level innovation;
        marketing innovation;
        organisational innovation;
        process innovations;
        product innovations
business-like surveys 91, 93

Canada 23, 30
Canberra II Group 60
Caribbean 47, 54
capital investment 60
Chaminade, C. 11, 26
Charmes, J. 70
civil society 29, 34
climate change 79, 85, 86, 89
Cluster Observatory 24
cluster policy 24
Colecchia, A. 100
collaboration 32–3, 34, 55, 74, 96
Commission of Experts for Research
    and Innovation 23
communities of practice 15–16, 37,
    55, 66, 71
Community Innovation Survey (CIS)
    9, 15, 30, 37, 43, 50, 61–2, 70

Compendium of Evidence on the
    Effectiveness of Innovation
    Policy Intervention Project 24
competence building 26
complexity 7, 8–9, 11–12
'connection between the user and the
    producer' 53, 62, 73, 105
consumer innovation 66
cooperation 55, 65
coordination 84, 89
country innovation policies 24
country innovation reports 10, 29,
    30–31
country innovation reviews 29, 31, 35
country ranking based on innovation
    indicators 30, 31–2, 34, 36
Crease, R.R. 17
*Creating Healthy Lives – The Future
    of Innovation* 35
cyber security 89
cybercrime 62

DARPA 41
data collection 64, 87–8, 92–3
Dear Colleague Letter 17
definitions of innovation
    evolution of, outside the business
        sector 36–7
    Oslo Manual see Oslo Manual
    social innovation 75, 76
'Delayed transfers of care' statistics
    75
demand-side innovation policies 44
demography, and innovation policy
    22
desirable outcomes 3
developing countries 4, 22, 26, 49,
    54, 99
developing innovation policy *see*
    innovation policy(ies),
    developing
digital economy 2–3
    agenda for action 104–5
    data collection 92, 93
    and innovation 12, 77, 95
    innovation policy 94–6, 98, 99

measurement of innovation 53–4,
        56, 91, 94–6
    OECD 2017 Scoreboard 32, 33
    as a priority for OECD
        innovation strategy 18
    in RIO – H2020 PSF 25
    services 96
    trends influencing 96
    zero price products 53, 54, 62,
        73, 98
Digital Economy Outlook (OECD) 25
Digital and Open Innovation Project
    (OECD) 44
digital platforms 99
digital transformation 94
digitalisation 32, 45, 56, 73, 95
direct intervention 41
direct measurement 29, 32, 33, 34,
    84, 85
Drummond, D. 23

e-commerce 73
Ebola virus 88
eco-innovation 24, 48, 85–7, 89–90,
    104
    definition 72
    measurement 71–2
economic sectors
    measurement of innovation
        across economic sectors
        69–77
    in all economic sectors
        58–67
    recognition of innovation as
        occurring in all 48
economically significant prices 53,
    61, 62
Edler, J. 41, 42
Edquist, C. 10, 11, 42
education 21–2
EFI 2019 report 23
employment surveys 70
Energising the World with Innovation
    35
Enhanced European Innovation
    Council (EIC) pilot 45
Ergas, H. 41

ethics 52, 75, 98
European Commission
    country information 24
    definition of social innovation 75
    promotion of green growth 86
    provision of approach to policy
        44–5
European Forum for the Studies
    of Policies of Research and
    Innovation (EUSPRI) 17
European Innovation Council 19
European Innovation Scoreboard
    (2019) 19, 33–4
European Public Sector Innovation
    Scoreboard (EPSIS) 36, 64
European Union (EU) 16
    CIS *see* Community Innovation
        Surveys
    eco-innovation 71
    Horizon 2020 framework
        programme 19, 45
    innovation performance groups
        33
    innovation policy 18–19
    Innovation Policy Fact Sheet 17,
        18, 99
    Innovation Union 19
    SDG targets 82
    setting standards for the digital
        economy 95
    success of *Oslo Manual* 103
Eurostat 10, 30, 54, 61, 65, 103
expenditure on R&D
    and innovation 30–31
    permission to claim as tax credit
        22–3
    propensity to innovate 20

Fagerberg, J. 41, 42
Federal Ministry of Economic Affairs
    and Energy (Germany) 24
feedback loops 4, 7, 98
finance ministries 100
financial corporations 61
financial crisis (2008), and innovation
    policy 40–41
financial support, start-ups 23–4

Finland 31, 37, 66
firm-level innovation 19
    green growth 86
    guidance on measuring 47–8
    OECD 2017 Scoreboard 33
    R&D and 20, 30–31
    surveys 92
    see also growth of firms; size of
        firms
fourth industrial revolution 40, 72–3
fragmented innovation policy 25, 50
framework conditions 7, 8, 26, 29, 32,
    35, 41, 57, 86, 96, 98
Framework for the Development
    of Environmental Statistics
    (FDES) 79, 85, 86
Frascati Manual (OECD, 2002) 54
Frascati Manual (OECD, 2015) 37,
    55, 104
Freeman, C. 16
funding programmes (Germany) 24
future-related policy 24

Gault, F. 12, 36, 37, 42, 52, 63
general definition of innovation 3–4,
    48, 49, 91, 103
    applicability in all sectors 50
    eco-innovation 72
    household sector 66
    informal sector 70
    measurement of presence of
        innovation 77
    in monitoring and evaluating
        policy 89
    NPISH sector 65
    product innovation 53
    public sector innovation 63
    see also 'made available to
        potential users'; 'new
        or improved product or
        process'
general government sector,
    measurement of innovation
    63–5
general purpose technologies and
    practices 72–3, 104
    policy 74

process innovation and user
    innovation 73–4
Germany 22, 24
Global Innovation Index (GII) 35
Global Observatory of Science,
    Technology and Innovation
    Policy Instruments (GO-SPIN)
    31
Global Partnership for Sustainable
    Development 85
global value chains 26, 67, 99, 101
Going Digital: Making the
    Transformation Work for
    Growth and Wellbeing
    (OECD) 94
Goldmann, M. 36
governance, and innovation policy
    40, 42
green economy 71
green growth 79, 85–6, 89–90
Green Growth Strategy (OECD) 85,
    89
green innovation *see* eco-innovation
green.eu project 71
growth of firms, R&D, innovation and
    promotion of 21

*Handbook of Innovation in Public
    Services* 36
Haskel, J. 60
health 22, 48, 79, 82, 83, 87, 88–9, 90
High-Tech Start-Up Fund 24
Hill, C.T. 42
holistic innovation policy 25, 42, 97
Hollanders, H. 32
Horizon 2020 programme (EU) 19, 45
Horst, J. 99
household innovation 3
    measurement 37, 65–6
    surveys 91, 93
household sector 7, 37, 65, 66, 68, 72,
    91, 93, 97, 102, 103
Huber, D. 64

Iceland 30
implementation of innovation 50, 51

implemented innovation policy
    12–13, 16, 18
inclusive growth 87–8, 90
inclusive innovation 44, 48, 52
indicators
    of welfare 94–5
    see also innovation indicators
indirect intervention 41
industrial classifications 92, 93, 100,
    101
Industry 4.0 99
InfoBrief 30
informal economy
    innovation in 4, 15, 26, 104
    measurement of innovation 54,
        69–70
information and communication
    technologies (ICT)
    manufacturing 33
information technology services 33
Innobarometer 19
innovation
    absence of official statistics on
        67
    confirmation of presence of 58
    in the context of the SciSIP
        project 16
    definitions *see* definitions of
        innovation
    digital economy and 12
    is everywhere 3
    future challenges 79–90
        climate change 85
        green growth and
            eco-innovation 85–7
        health issues 88–9
        inclusive growth 87–8
        SDGs and innovation
            measurement 80–85
    influence of standards 55
    in the informal economy 4, 15,
        26
    leaders 33
    measurement *see* statistical
        measurement
    policy *see* innovation policy(ies)
    positive environment for 18

priorities for 18
recognition of occurrence in all
    sectors 48
research *see* research
see also business sector
    innovation;
    eco-innovation; household
    sector innovation; public
    sector innovation; social
    innovation
innovation activities 11
in the digital economy 95
household sector 66
inclusive growth 88
Innovation in Firms (OECD 2017
    Scoreboard) 33
innovation indicators
country ranking based on 30,
    31–2, 36
digital economy 94
publication 10
science, technology and
    innovation (STI) 16
for showing outcomes of policy 3
sustainable development goals
    82, 83–4, 85
use of 9
innovation policy(ies) 4
in the 2020s
    digital economy 94–6, 98,
        99
    networks 98–9
    understanding and acting on
        99–100
challenge of producing indicators
    for showing outcomes 3
developing 40–45
    governance 40, 42
    international policy 44–5
    measurement and
        restrictions 43–4
    OECD working examples 44
    operational perspective 42
    time as an important variable
        in 22
implemented 12–13, 16, 18
inclusive 87, 88

influences on
    country ranking 34
    general purpose technologies
        74
    measurement of innovation
        2
monitoring and evaluation 29–38,
    42, 89
    beyond the business sector
        36–8
    country reports 29, 30–31
    country reviews 29, 31, 35
    importance 103
    ranking 30, 36
    scoreboards 29–30, 31–6
    trial and error 87
prior to 2020
    in countries 24–5
    EU innovation policy 18–19
    history 15–17
    international and
        supranational
        organisations 17
    national innovation systems
        25–6
    objectives 21–4
    OECD and innovation
        strategy 17–18
    statistical surveys 19–21
    studying innovation systems to
        support development of 97
    systems approach 19, 41, 89
    undeveloped theoretical basis for
        designing 10
Innovation Policy Fact Sheet (EU) 17,
    18, 99
Innovation Policy Platform (IPP) 24
Innovation Strategy (OECD, 2010) 18
Innovation Strategy (OECD, 2015)
    17–18, 42
innovation systems 9–10
    complexity and non-linearity 8–9
    literature 10–11
    see also national innovation
        systems
Innovation Union 19

innovative activities, social innovation 75, 76
Institute for Statistics (UNESCO) 54, 95
institutional units
    business sector 49
    definition 7–8, 59
    digital platforms 99
    government sector 63
    household sector 65
    people-related innovation 82
    registers 93
Intergovernmental Meeting on Science, Technology and Innovation Indicators 54
international health issues 79
international innovation policy 44–5
International Labour Organization (ILO) 70
International Monetary Fund (IMF) 94–5
international organisations
    cooperation 55
    country reviews 31
    guidance on standards for digital economy 95
    publishing of information on innovation policy 17
    see also individual organisations
International Organization for Standardization (ISO) 100
    56000 series 49, 55
international scoreboards *see* scoreboards
International Standard Industrial Classification (ISIC) 92
international standards 55, 57
    Oslo Manual definitions as 4, 54–5, 57
International Statistical Institute (ISI) 95, 100
International Telecommunication Union (ITU) 100
interventions 8, 9, 41, 75, 89, 98
'introduced on the market' 16, 32, 37, 43, 50, 51, 55, 56, 61

invention-oriented innovation policy 41

Janz, N. 32

Kazakhstan 31
Kenya 31
knowledge generation 19, 22–3
knowledge transfer 19, 23–4
Kuusisto, J. 37
Kyoto Protocol 85

Latin America 47, 54
leaders of innovation 33
learning 26
linkages 4, 7, 8, 26, 47, 67, 87, 97–8
Lundvall, B.-Å. 26
Lurtz, K. 75

Maastricht Manual on Measuring Eco-Innovation for a Green Economy 71, 72, 86, 87, 89–90
MacCallum, D. 75
'made available to potential users' 3, 37, 49, 51, 53, 56, 62, 66, 70, 76, 82, 105
malaria treatment 88–9
Manchester Institute of Innovation Research (MIoIR) 17, 24
manuals
    internationally agreed 100
    for measurement of public sector innovation 64–5
    see also individual manuals
manufacturing 33, 47, 48, 76
Marburger, J. 15, 16, 17, 42, 100
Marée, M. 76
market, role in innovation 61–2
market economy 95
marketing innovation 47, 50, 51
Mashelkar, R.A. 52
Meadows, D. 12
measurability of innovation 33
measurement of innovation *see* statistical measurement

Measuring Public Innovation in the
    Nordic Countries (MEPIN)
    project 36–7, 63, 64
Mertens, S. 76
methodological constraints 43
Millennium Development Goals
    (MDGs) 82
ministers of research/education/
    technology/innovation 100
mission-oriented innovation policy 41
moderate innovators 33
modest innovators 34
Molotja, N. 23, 30
monitoring 4, 9, 29–40, 70–76, 79,
    80, 82, 87, 88, 99, 103
Moulaert, F. 75
Mulgan, G. 75, 76
Murdoch, A. 75
Murray, R. 76

National Academies of Sciences,
    Engineering and Medicine 93
National Center for Science and
    Engineering Statistics (NCES)
    93
National Endowment for Science,
    Technology and the Arts
    (Nesta) 24
national innovation systems
    definition 26
    innovation policy 25–6
    literature 11–12
National Research Council 93
National Science Foundation (NSF)
    15–16, 17, 19, 93
networks 4, 34, 63, 98–99
'new or improved product or process'
    3, 16, 37, 43, 49, 50, 53, 55,
    62, 65–6, 72, 76, 105
New Partnership for Africa's
    Development (NEPAD) 54
Nicholls, A. 75
non-economically significant prices
    62, 67, 70
non-financial corporations 61
non-linearity (system) 8–9

non-profit institutions serving
    households (NPISHs) 65, 91,
    93
Norway 30, 31, 34
NPISH, definition 7

Observatory of Public Sector
    Innovation (OPSI) 36, 63–4
OECD
    -World Bank Innovation Policy
        Platform 24
    Blue Sky Forum I (1996) 48
    Blue Sky Forum II (2006) 15, 16,
        42, 100
    country innovation reviews 29,
        31
    country profiles 24
    Digital Economy Outlook 25
    Digital and Open Innovation
        Project 44
    Frascati Manual (2002) 54
    Frascati Manual (2015) 37
    Going Digital: Making the
        Transformation Work for
        Growth and Wellbeing 94
    Green Growth Strategy 85, 89
    inclusive innovation policies 87
    Innovation Strategy (2010) 18
    Innovation Strategy (2015)
        17–18, 42
    international standards
        maintenance 54
    member countries 16, 87
    Observatory of Public Sector
        Innovation (OPSI) 36,
        63–4
    R&D Tax Incentives Database
        22–3
    Science, Technology and
        Industry Scoreboard
        (2017) 32–3
    setting standards for the digital
        economy 95
    success of *Oslo Manual* 103
    work on clusters 24
    working examples of innovation
        policy 44

Working Party of National
Experts on Science and
Technology Indicators
(NESTI) 10, 37, 54, 60,
64–5, 71, 72, 87, 103–104
*OECD Science, Technology and
Industry Outlook 2014*
(OECD) 24
off-the-shelf technologies 74
official statistics 91
absence of 67
agenda for action 103–4
importance 71
inclusion of data on use
and planned use of
technologies 96
statistical offices and production
of 10
open innovation, EU interest in 19
organisational innovation 47, 50, 51
Oslo Manual 65
Annex interpreting the manual
(third edition) for use in
developing countries 47,
49, 57
definitions of innovation
business sector *see* business
sector innovation
eco-innovation 72
evolution of 9
household innovation 37
as international standards 4,
54–5, 57
for measurement purposes
3–4, 47–8, 49–51
see also general definition of
innovation
first edition (1992) 47, 48, 59,
71, 90
fourth edition (2018) 3, 4, 36, 38,
43, 48, 49, 50–51, 54, 59,
61, 66, 76, 102
global accessibility 47
presence of innovation
'everywhere' 102
second edition (1997) 47, 48
success 103–4

third edition (2005) 11, 16, 37,
43, 47, 48, 50–51, 54, 102
understanding of 'value' 55
Oxford Handbook of Innovation
10–11

Paris Agreement 85
partnership 81, 84–5
Paunov, C. 12, 88, 95, 96, 98
peace 81, 84
people
innovation related to 21–2
SDGs and innovation
measurement 80, 82
Planes-Satorra, S. 12, 88, 95, 96, 98
planet 80–81, 83
platforms 2, 99
policy learning 4, 9, 24, 25, 42, 62,
87, 89
Portugal 29, 31
positive environment, for innovation
18
potential users 68
see also 'made available to
potential users'
poverty 82
priorities for innovation 18
pro-poor innovation 48
process innovations
general purpose technologies
73–4
household sector 65–6
Oslo Manual definitions 16, 47,
50–51
see also 'brought into use';
'new or improved
product or process'
R&D and 20
restrictions on measurement of
44
producers, connection between users
and 53, 62, 73, 105
product innovations 3
education as a policy objective
21
general purpose technologies 73
household sector 65–6

informal sector 70
Oslo Manual definitions 47
see also 'introduced on the market'; made available to potential users; 'new or improved product or process'
R&D and 20
restrictions on measurement of 44
zero price product 54, 62, 73, 104
propensity to innovate 20, 30, 48
prosperity 81, 83–4
prototypes 66
proximity 24
public sector 7–8
public sector innovation 41, 84
definitions 36–7, 59–60
EC provision of approach to 44
EU support for 19
interest in 15
measurement 38, 63–5
studies 63–4
surveys 93

R&D
business sector performers 23
and innovation 20, 21, 30–31
as innovation policy 4, 18, 22–3
invention-oriented innovation policy 41
SNA attempt to capitalise 60, 68
see also expenditure on R&D
R&D Tax Incentives Database 22–3
Ramlogan, R. 24
Rammer, C. 23
ranking see country ranking based on innovation indicators
regional development policy 40
Regional Innovation Scoreboard (RIS) 19, 34
research
and development see R&D
on innovation 15

integration of innovation and 19
Research Excellence and Collaboration (OECD 2017 Scoreboard) 32–3
Research and Innovation (electronic platform) 25
Research and Innovation Observatory – Horizon 2020 Policy Support Facility (RIO – H2020 PSF) 25
research institutes 10, 17, 23, 71, 90
residential institutional units 59
Rest of the World (ROW) 67
restricted innovation 3, 4, 43–4, 48, 51–3, 56, 102
see also eco-innovation; inclusive innovation; sustainable innovation
Revision 4 (ISIC) 92
RICYT (Iberoamerican Network of Science and Technology) 47, 54
role of the market in innovation 61–2
role of policy makers 19
Romania 34
Rüede, D. 75

Santiago, F. 99
Schubert, T. 23
Science and Engineering Indicators 2018 30
Science Policy Research Institute (University of Sussex) 17
Science of Science: Discovery, Communication, Impact (SoS: DCI) Program 17
Science of Science and Innovation Policy (SciSIP) 15–16, 17
Science of Science Policy 15
science, technology, engineering and mathematics (STEM) 21
Science, Technology and Industry Scoreboard (OECD, 2017) 32–3
science, technology and innovation (STI) 16, 25, 100

see also Working Party of
National Experts on
Science and Technology
Indicators (NESTI)
Science, Technology and Innovation
Strategy for Africa (STISA –
2024) 45
Scientific Research and Experimental
Development (SR&ED) tax
benefit programme 23
scoreboards 29–30
European Innovation Scoreboard
(2019) 19, 33–4
as a means of ranking 31–2
OECD Science, Technology
and Industry Scoreboard
(2017) 32–3
Regional Innovation Scoreboard
(RIS) 19, 34
use of indicators in 9
sector-specific definition 49, 58, 69
sectoral innovation policy 40
Serbia 34
size of firms
and collaboration 32–3
influence on innovation 33
public support for innovation 33
R&D and innovation 20, 21
small and medium-sized enterprises
(SMEs) 23, 32–3, 34
smallpox virus 88
Smith, K. 61
SNA Manual 2008 60, 63
social innovation 3, 104
definitions 75, 76
EC provision of approach to 44
EU support for 19
interest in 15
lack of internationally supported
definition 4
measurement 75–7
Social Innovation Blog of the Social
Innovation Academy 75
social surveys 91, 93
Soete, L. 16
South Africa 23

standard industrial classifications 92,
93, 100, 101
start-ups, financial support 23–4
statistical measurement
across economic sectors 69–77
eco-innovation and green
growth 71–2, 86
general purpose technologies
72–4
informal economy 69–70
social innovation 75–7
agenda for action 103–5
in all economic sectors 58–67
business sector 42, 61–3
general government sector
and public sector 38,
63–5
household sector 37, 65–6
non-profit institutions
serving households
(NPISHs) 65
the Rest of the World
(ROW) 66
the System of National
Accounts 58–60
in a changing world 2–3
climate change 85
data collection 92–3
defining innovation for 10, 47–57
digital economy 53–4, 56
Oslo Manual 3–4, 49–51,
54–5
restricted innovation 51–3,
56
desirable outcomes and need for
restrictions on 3
digital economy 53–4, 56, 91,
94–6
first formalised guidelines 15
inclusive growth 87
influence on policy 2
internationally agreed manuals
100
sustainable development goals
82–5
systems approach 47, 48, 96–7
see also official statistics

STIP Compass on International
    Database on STI policies 25
strong innovators 33
supranational organisations 17, 95
    see also African Union (AU);
        European Union (EU);
        OECD
Survey on Science, Technology and
    Innovation Policy (STIP) 24–5,
    27
surveys 9, 30, 91
    business sector innovation 19–21
    firms 92
    informal sector 70
    methodological constraints 43
    public sector innovation 64
    see also Community Innovation
        Survey (CIS)
sustainable development 79, 80
Sustainable Development Goals
    (SDGs) 24, 79, 80–81
    appearance of word 'innovation'
        in 81–2
    innovation measurement 82–5,
        86
    targets 82
    use of innovation policy 89
sustainable innovation 44, 48, 52
Switzerland 34
System of National Accounts (SNA)
    7–8, 49, 58–60, 102
system-oriented policy 41
systems
    boundedness 7, 8
    components 7–8
    intervention, complexity and
        non-linearity 8–9
    see also innovation systems
systems analysis 8, 11–12
systems approach 12
    innovation policy 19, 41, 89
    measurement of innovation 47,
        48, 96–7

tax credits 22–3
Teich, A.H. 16, 27

transformation, digital 94
three-year reference period, in CIS 43
time
    in innovation policy development
        22
    in public sector innovation 64
    in restricted innovation 52
    and understanding of the digital
        economy 96
    'to guide discussion and theorizing'
        53, 69, 76

UN Framework Convention on
    Climate Change 85
UN Industrial Development
    Organization (UNIDO) 31
UN Statistical Commission 92
UN Statistics Division (UNSD) 85,
    92, 95
UN Sustainable Development Goals
    24
understanding innovation policy
    99–100
UNESCO 31, 54, 95
United Kingdom 17, 30, 75, 82
United States 16, 30, 42, 93
user innovation 74, 96
users
    connection between producers
        and 53, 62, 73, 105
    see also potential users
Uyarra, E. 24

value 55
value chains 26, 47, 67, 99, 101
von Bogdandy, A. 36
von Hippel, E. 37
vouchers 23, 25

welfare, indicators of 94–5
Westlake, S. 60
'whole of government approach' 25,
    40, 42
Wiesbaden Group on Business
    Registers 92

Working Party of National Experts
     on Science and Technology
     Indicators (NESTI) 10, 37, 54,
     60, 64–5, 71, 72, 87, 103–4
workplace innovation 44
World Bank 24, 31, 95
World Intellectual Property
     Organisation (WIPO) 95, 100

Wright, D. 12

young population, engagement of 22

zero price products 53, 54, 62, 70,
     73, 98